Studies in Big Data

Volume 70

Series Editor

Janusz Kacprzyk, Polish Academy of Sciences, Warsaw, Poland

The series "Studies in Big Data" (SBD) publishes new developments and advances in the various areas of Big Data- quickly and with a high quality. The intent is to cover the theory, research, development, and applications of Big Data, as embedded in the fields of engineering, computer science, physics, economics and life sciences. The books of the series refer to the analysis and understanding of large, complex, and/or distributed data sets generated from recent digital sources coming from sensors or other physical instruments as well as simulations, crowd sourcing, social networks or other internet transactions, such as emails or video click streams and other. The series contains monographs, lecture notes and edited volumes in Big Data spanning the areas of computational intelligence including neural networks, evolutionary computation, soft computing, fuzzy systems, as well as artificial intelligence, data mining, modern statistics and Operations research, as well as self-organizing systems. Of particular value to both the contributors and the readership are the short publication timeframe and the world-wide distribution, which enable both wide and rapid dissemination of research output.

** Indexing: The books of this series are submitted to ISI Web of Science, DBLP, Ulrichs, MathSciNet, Current Mathematical Publications, Mathematical Reviews, Zentralblatt Math: MetaPress and Springerlink.

More information about this series at http://www.springer.com/series/11970

Jorge Vázquez-Herrero · Sabela Direito-Rebollal ·
Alba Silva-Rodríguez · Xosé López-García
Editors

Journalistic Metamorphosis

Media Transformation in the Digital Age

 Springer

Editors
Jorge Vázquez-Herrero
Faculty of Communication Sciences
Universidade de Santiago de Compostela
Santiago de Compostela, Spain

Sabela Direito-Rebollal
Faculty of Communication Sciences
Universidade de Santiago de Compostela
Santiago de Compostela, Spain

Alba Silva-Rodríguez
Faculty of Communication Sciences
Universidade de Santiago de Compostela
Santiago de Compostela, Spain

Xosé López-García
Faculty of Communication Sciences
Universidade de Santiago de Compostela
Santiago de Compostela, Spain

ISSN 2197-6503 ISSN 2197-6511 (electronic)
Studies in Big Data
ISBN 978-3-030-36317-8 ISBN 978-3-030-36315-4 (eBook)
https://doi.org/10.1007/978-3-030-36315-4

This Springer imprint is published by the registered company Springer Nature Switzerland AG
The registered company address is: Gewerbestrasse 11, 6330 Cham, Switzerland

Editorial Project

This book is part of the activities developed in:

- the research project *Digital native media in Spain: storytelling formats and mobile strategy* (RTI2018-093346-B-C33) funded by the Ministry of Science, Innovation and Universities (Government of Spain), Agencia Estatal de Investigación, and co-funded by the European Regional Development Fund (ERDF);
- the research project *New values, governance, funding and public media services for the Internet society: European and Spanish contrasts* (RTI2018-096065-B-I00) funded by the Ministry of Science, Innovation and Universities (Government of Spain), Agencia Estatal de Investigación, and co-funded by the European Regional Development Fund (ERDF); and
- Novos Medios research group, supported by Xunta de Galicia (ED431B 2017/48).

Editorial coordination, book layout and formatting: Jorge Vázquez-Herrero, Sabela Direito-Rebollal and Alba Silva-Rodríguez.

Preface

Over the past decades, journalism has been in a state of flux due to the outcome of the digitalization of the media environment. The increase of channels and platforms together with the emergence of digital native players has created an immediate need for journalism to innovate in order to survive in this digital and social media age. However, although innovation has turned into a significant matter to guarantee the future of journalism, it is still necessary to deeply understand its potential implications within the media industry. In order to identify relevant features for media innovation, this book emphasizes three dimensions of change: content and narrative, technology and formats, and business models.

Journalistic Metamorphosis: Media Transformation in the Digital Age is divided into four parts. The first part consists of three chapters which cover the technological impact of the challenges and consequences in media. Ramón Salaverría and Mathias-Felipe de-Lima-Santos (University of Navarra) study the implementation of the so-called Internet of things (IoT) within the framework of technological innovations assimilated by journalism over the last 25 years. They describe devices, applications and systems that media have incorporated in the production and consumption of news content, by providing a general overview of the opportunities and challenges that IoT offers to journalism.

The debate about automated journalism is addressed in the chapter signed by José Miguel Túñez-López, Carlos Toural-Bran (Universidade de Santiago de Compostela) and Ana Gabriela Frazão-Nogueira (Universidade Fernando Pessoa). The automation of the search engines, classification and the treatment of information as a result of the alteration of a journalist's routine due to the consequences of the implementation of AI to the workflow is part of the analysis made by the aforementioned authors.

The search for new ways of journalistic narratives is the main topic of the chapter by Jorge Vázquez-Herrero, Xosé López-García (Universidade de Santiago de Compostela) and Fernando Irigaray (Universidad Nacional de Rosario), for whom ubiquity, transmediality and micro-contents define emerging narratives, with a strong link with social media and their consumption.

The second part of this book revolves around these emerging narratives and journalistic formats. Ana Cecília B. Nunes (Pontifical Catholic University of Rio Grande do Sul) and João Canavilhas (University of Beira Interior) look for the identification of peculiarities in innovative journalism taking as a reference the case study of Google DNI Fund initiatives in the European context highlighted in its three-year report. Digital revenue and news narratives or formats are the main goals to achieve consumers' needs. Jonathan Hendrickx, Karen Donders and Ike Picone (Vrije Universiteit Brussel) discuss the unexpected popularity of e-mail newsletters as a successful tool for legacy and new news media outlets alike to independently disseminate their content, without having to rely on social media algorithms. Andreu Casero-Ripollés, Silvia Marcos-García and Laura Alonso-Muñoz (Universitat Jaume I) present new social media formats for local journalism like live blogging, summary information through WhatsApp or Instagram Stories and 360° image and video.

The continuous rise of visual information available online is prompted by the current state of the so-called liquid modernity. Ángel Vizoso (Universidade de Santiago de Compostela) focuses his work on this topic by underlining the weak and strong points of this visual system. Sara Pérez-Seijo and Berta García-Orosa (Universidade de Santiago de Compostela) analyze the use of this immersive narrative in five international NGOs from an ethical and critical point of view.

The third part of this book, which begins with the one signed by Ainara Larrondo Ureta, Koldo Meso Ayerdi and Simón Peña Fernández (University of the Basque Country), addresses the impact that social networking sites have caused in the legacy media and its continuous use as a way to innovatively enhance consumer engagement. The authors show a global vision of the strategies used by Spanish media concerning social networks, especially regarding the development of professional practices adapted to new patterns of users' consumption, as well as new modes of content creation, distribution and promotion.

News organizations publish social media guidelines to guide the conduct of their journalists on social networks, due to the common use of Facebook and Twitter to collect, distribute, promote and discuss the news. Sabela Direito-Rebollal, María-Cruz Negreira-Rey and Ana-Isabel Rodríguez-Vázquez (Universidade de Santiago de Compostela) study the social media guidelines developed by public service broadcasting corporations in the European Union to explain the recommendations for their journalists.

The last two chapters of the third segment are focused on the innovation of broadcasting formats. Jose A. García-Avilés (Miguel Hernández University) analyzes six new international broadcasting formats. The author concludes that innovation in newscasts essentially lies in the audiovisual narrative, integrating image and sound, telling the story with fluency and holding viewers' trust with rigorous journalism. This topic is also considered by Ana González-Neira and Natalia Quintas-Froufe (Universidade da Coruña) with their analysis of TV's audience participation in news bulletins.

Last part of this book puts together three chapters that delve into the study of digital media revenue business models and in the era of the so-called post-journalism, focusing on the figure of the journalist, particularly in its loss of legitimacy.

Manuel Goyanes (Universidad Carlos III de Madrid), Marta Rodríguez-Castro and Francisco Campos-Freire (Universidade de Santiago de Compostela) analyze the relation between innovation and digital media revenue business models. The authors consider that the core relies on social value and creative intelligence. Laura Solito and Carlo Sorrentino (Università degli Studi di Firenze) study the practices journalists use to authenticate their work. They argue that journalistic authority is always the product of complex and variable relationships. Furthermore, the authors go in depth studying the changes of the professional identity in consequence of the loss of trust from the audience and the difficulties encountered from traditional business models within the industry. Finally, Xosé López-García, Alba Silva-Rodríguez, Sabela Direito-Rebollal and Jorge Vázquez-Herrero (Universidade de Santiago de Compostela) theorize about the future role of journalists in the era of post-journalism. The authors come to the conclusion that journalism is not dying, but it is returning to its roots. In order to do so, it is necessary to implement data journalism, immersive journalism, multimedia and multi-format narratives, trans-media narratives, verification techniques (fact-checking) or semi-automated systems.

Journalistic Metamorphosis: Media Transformation in the Digital Age, edited by Jorge Vázquez-Herrero, Sabela Direito-Rebollal, Alba Silva-Rodríguez and Xosé López-García (Universidade de Santiago de Compostela), brings together experts from Europe and America working on the impact of technology, the reconfiguration of the media ecosystem and the transformation of business models within the context of glocal information and enriched innovation.

Santiago de Compostela, Spain

Jorge Vázquez-Herrero
Sabela Direito-Rebollal
Alba Silva-Rodríguez
Xosé López-García

Contents

Towards Ubiquitous Journalism: Impacts of IoT on News   1
Ramón Salaverría and Mathias-Felipe de-Lima-Santos

From Data Journalism to Robotic Journalism: The Automation
of News Processing   17
José Miguel Túñez-López, Carlos Toural-Bran
and Ana Gabriela Frazão-Nogueira

The Technology-Led Narrative Turn   29
Jorge Vázquez-Herrero, Xosé López-García and Fernando Irigaray

Journalism Innovation and Its Influences in the Future of News:
A European Perspective Around Google DNI Fund Initiatives   41
Ana Cecília B. Nunes and João Canavilhas

Innovating Journalism by Going Back in Time? The Curious
Case of Newsletters as a News Source in Belgium   57
Jonathan Hendrickx, Karen Donders and Ike Picone

New Formats for Local Journalism in the Era of Social Media
and Big Data: From Transmedia to Storytelling   69
Andreu Casero-Ripollés, Silvia Marcos-García and Laura Alonso-Muñoz

Information Visualization and Usability: Tools for Human
Comprehension   85
Ángel Vizoso

Use of 360-Degree Video in Organizational Communication:
Case Study of Humanitarian Aid NGOs   99
Sara Pérez-Seijo and Berta García-Orosa

Shared Spaces for News Content Production in Spanish
Online Media 113
Ainara Larrondo Ureta, Koldo Meso Ayerdi and Simón Peña Fernández

**Social Media Guidelines for Journalists in European Public
Service Media** 129
Sabela Direito-Rebollal, María-Cruz Negreira-Rey
and Ana-Isabel Rodríguez-Vázquez

**Reinventing Television News: Innovative Formats in a Social Media
Environment** 143
Jose A. García-Avilés

**Mediamorphosis of Participation on Television: The News
Programmes** 157
Ana González-Neira and Natalia Quintas-Froufe

Value and Intelligence of Business Models in Journalism 171
Manuel Goyanes, Marta Rodríguez-Castro and Francisco Campos-Freire

New Forms of Journalistic Legitimization in the Digital World 185
Laura Solito and Carlo Sorrentino

From Meta-Journalism and Post-Journalism to Total Journalism 199
Xosé López-García, Alba Silva-Rodríguez, Sabela Direito-Rebollal
and Jorge Vázquez-Herrero

Editors and Contributors

About the Editors

Jorge Vázquez-Herrero is Ph.D. in Communication, Universidade de Santiago de Compostela (USC). He is Member of Novos Medios research group (USC) and Latin-American Chair of Transmedia Narratives (ICLA–UNR, Argentina). He was Visiting Scholar at Universidad Nacional de Rosario, Universidade do Minho, University of Leeds and Tampere University. His research focuses on digital interactive non-fiction digital storytelling—mainly interactive documentary—micro-formats and transmedia, immersive and interactive narratives in online media.

Sabela Direito-Rebollal is Ph.D. Candidate at Universidade de Santiago de Compostela (USC). She holds a Degree in Audiovisual Communication (USC), a Master's Degree in Communication and Creative Industries (USC) and a Degree in Movie and TV Script from the Madrid Film Institute. She was visiting scholar at the University of Hull (United Kingdom) and the Vrije Universiteit Brussel (Belgium). Her research focuses on innovation, audience trends and programming strategies of European public service media.

Alba Silva-Rodríguez is Associate Professor of journalism at the Department of Communication Sciences at Universidade de Santiago de Compostela (USC). She is Ph.D. in journalism and Member of Novos Medios research group. She is Secretary of the RAEIC journal. As Researcher, she focuses on the assessment of digital communication, especially the study of mediated conversation in social media and the evolution of media contents in mobiles devices.

Xosé López-García is Professor of journalism at Universidade de Santiago de Compostela (USC) and a Ph.D. candidate in history and journalism (USC).

He coordinates the Novos Medios research group. Among his research lines, there is the study of digital and printed media, analysis of the impact of technology in mediated communication, analysis of the performance of cultural industries and the combined strategy of printed and online products in the society of knowledge.

Contributors

Laura Alonso-Muñoz Universitat Jaume I, Castellón, Spain

Francisco Campos-Freire Universidade de Santiago de Compostela, Santiago de Compostela, Spain

João Canavilhas University of Beira Interior, Covilhã, Portugal

Andreu Casero-Ripollés Universitat Jaume I, Castellón, Spain

Mathias-Felipe de-Lima-Santos University of Navarra, Pamplona, Spain

Sabela Direito-Rebollal Universidade de Santiago de Compostela, Santiago de Compostela, Spain

Karen Donders Vrije Universiteit Brussel, Brussels, Belgium

Ana Gabriela Frazão-Nogueira Universidade Fernando Pessoa, Porto, Portugal

Jose A. García-Avilés Miguel Hernández University, Elche, Spain

Berta García-Orosa Universidade de Santiago de Compostela, Santiago de Compostela, Spain

Ana González-Neira Department of Sociology and Communication Sciences, Universidade Da Coruña, A Coruña, Spain

Manuel Goyanes Universidad Carlos III de Madrid, Madrid, Spain

Jonathan Hendrickx Vrije Universiteit Brussel, Brussels, Belgium

Fernando Irigaray Universidad Nacional de Rosario, Rosario, Argentina

Ainara Larrondo Ureta University of the Basque Country, Leioa, Spain

Xosé López-García Universidade de Santiago de Compostela, Santiago de Compostela, Spain

Silvia Marcos-García Universitat Jaume I, Castellón, Spain

Koldo Meso Ayerdi University of the Basque Country, Leioa, Spain

Ana Cecília B. Nunes Pontifical Catholic University of Rio Grande do Sul, Porto Alegre, Brazil

María-Cruz Negreira-Rey Universidade de Santiago de Compostela, Santiago de Compostela, Spain

Ike Picone Vrije Universiteit Brussel, Brussels, Belgium

Sara Pérez-Seijo Universidade de Santiago de Compostela, Santiago de Compostela, Spain

Simón Peña Fernández University of the Basque Country, Leioa, Spain

Natalia Quintas-Froufe Department of Sociology and Communication Sciences, Universidade Da Coruña, A Coruña, Spain

Marta Rodríguez-Castro Universidade de Santiago de Compostela, Santiago de Compostela, Spain

Ana-Isabel Rodríguez-Vázquez Universidade de Santiago de Compostela, Santiago de Compostela, Spain

Ramón Salaverría University of Navarra, Pamplona, Spain

Alba Silva-Rodríguez Universidade de Santiago de Compostela, Santiago de Compostela, Spain

Laura Solito Dipartimento di Scienze Politiche e Sociali, Università degli Studi di Firenze, Florence, Italy

Carlo Sorrentino Dipartimento di Scienze Politiche e Sociali, Università degli Studi di Firenze, Florence, Italy

Carlos Toural-Bran Universidade de Santiago de Compostela, Santiago de Compostela, Spain

José Miguel Túñez-López Universidade de Santiago de Compostela, Santiago de Compostela, Spain

Jorge Vázquez-Herrero Universidade de Santiago de Compostela, Santiago de Compostela, Spain

Ángel Vizoso Universidade de Santiago de Compostela, Santiago de Compostela, Spain

Towards Ubiquitous Journalism: Impacts of IoT on News

Ramón Salaverría and Mathias-Felipe de-Lima-Santos

Abstract After embracing web and mobile technologies, the news media are ready to receive a third technological wave: the Internet of Things (IoT). This set of technologies has already begun to spread, through new devices based on artificial intelligence. One of the most affected areas by this new technological wave will be that of journalistic information. Robotic systems and IoT devices are bringing new modes of production, distribution and consumption of the journalistic content, taking the news media to a new paradigm: the ubiquitous journalism. This chapter grounds on a historical perspective to place the implementation of the IoT within the framework of technological innovations assimilated by journalism over the last quarter of a century. It describes the devices, applications and systems that the media are being incorporated into the production and consumption of news content, providing a general overview of the opportunities and challenges that IoT poses to journalism.

Keywords Digital journalism · Internet of things · Robotization · News production

1 A Fish Tank Full of Information

Fish don't know they're in the water. They breathe, feed, move in that liquid, but they are not aware that they inhabit it and that, if they run out of water or it simply changes—in its degree of salinity, oxygenation or temperature—they won't survive.

This image of the close relationship between fish and water was often used half a century ago by Marshall McLuhan (Gossage 1967) to describe people's relationship with the news media. Today it remains as a good analogy, probably better than ever. In fact, recently other authors have kept comparing contemporary society (Bauman 2000) and even current journalism (Deuze 2008) with the liquid element. We live surrounded by an expanding set of communication technologies that connect us with

R. Salaverría (✉) · M.-F. de-Lima-Santos
University of Navarra, Pamplona, Spain
e-mail: rsalaver@unav.es

M.-F. de-Lima-Santos
e-mail: mdelimas@unav.es

© Springer Nature Switzerland AG 2020
J. Vázquez-Herrero et al. (eds.), *Journalistic Metamorphosis*,
Studies in Big Data 70, https://doi.org/10.1007/978-3-030-36315-4_1

other people and allow us to access all kinds of information and services. But, as if we were fish in a tank, they are increasingly transparent to us, almost imperceptible.

Information has become an ever-present good. From the very minute we wake up until we go to bed, all day long we cross over hundreds of messages that call for our attention: the time, the temperature, the traffic, the latest sports results, the evolution of stock market prices... And also, of course, all sort of news. An endless thread of news impacts, from simple textual flashes to highly elaborated multimedia packages, reach to our eyes and ears in the multiple screens and loudspeakers that we meet at each step. Even if we are not searching for news, they come to us, often creating an uncomfortable sense of information overload (Shenk 1997).

This is one of the great changes in the news media consumption patterns compared to previous eras. In the past, those who wished to access journalistic information had to make a deliberate act of searching. To read the content of a newspaper, they first had to go to a newspaper stand or buy a subscription. To watch or hear news on broadcast media, they had to attend news programs in scheduled times. Television and radio were devices located in a central place at home and the members of the family had to gather collectively in front of these almost totemic objects, at predefined times. They could not choose what content would be broadcasted by those devices; they could only decide whether they would consume or not what the audio-visual media companies offered. The audience was captive of programmers' decisions.

In the last quarter of a century, the digitalization of information has blown up those old rules of distribution of the commercial, journalistic and entertainment content. Today it is the public, each one of its members individually, who choose the place, the moment, the content and the format of what they want to consume. Or, at least, the public believes it has that power.

Although users can benefit from the increasing adaptability and advanced personalization of content, it is no less true that these contents begin to be shaped and targeted to each user inadvertently. This micro-targeting is possible thanks to artificial intelligence (AI) technologies, one of the fastest growing technological trends in contemporary journalism (Newman 2017). Thanks to the continuous use of digital devices and platforms by each user, these technologies allow the content providers to know every detail about the preferences and consumption patterns of each user. Thus, content providers and digital platforms are able to profile, with almost surgical precision, the interests of the users. These have the false impression that they control the flow of the content they consume, when, in fact, they are subject to automated decision-making by algorithmic systems. These AI technologies surreptitiously select and provide the journalistic, advertising and commercial contents that keep users' attention, following the interests of news providers and advertisers.

Without noticing it, contemporary media users—that is, all of us—live in a fish tank. And the liquid that surrounds us is called information.

1.1 News in the Age of IoT

The ever closer link between people and information is possible thanks to the evolution of the intermediary element between both of them: information technologies. In the last twenty-five years, the digitalization of technologies has transformed the content of the media, the profile of journalists, the news organizations and their audiences (Salaverría and Sádaba 2003). However, the way in which technologies continue to impact all those areas of journalism evolves over time. In the 1990s and during a large part of the first decade of the 21st century, digital technologies transformed mainly the news production processes, causing a profound disruption in aspects such as the professional profiles of journalists and media business models. The current technological evolution, on the other hand, is affecting mainly in other areas: it is transforming the public's relationship with information and the very notion of news.

Close to begin the third decade of the 21st century, to understand the future evolution of journalism, it is necessary to relate it to the Internet of Things (IoT) and all the technologies that it involves: real-time analytical systems of big data, self-improving algorithmic systems (also known as machine learning), interaction systems between sensors for data capture, monitoring tools and remote control of electronic objects, among others (Greengard 2015). They are applications that just a few years ago sounded like science fiction, but nowadays most of these IoT technologies are already a reality and are announced as hegemonic by the middle of the third decade of this century (Pew Research Center 2014). If in the early years of the Internet there was a process of digitalization of journalism, in the coming years the main technological transformation will be towards robotization and AI.

What impact will the IoT have on journalism? It is still early to know it. However, there is little doubt that this impact will be very large and diverse. The reason is that IoT technologies affect the three main phases of journalistic work: information gathering, processing and, especially, news content distribution.

With regard to the first phase, that of information gathering, journalism will be enriched by the enormous volume of information coming from the automatic sensors, present in almost any digital device. These sensors capture data in an uninterrupted way, transforming it into quantitative measures of all kind: from the temperature of the sea on the coast to the speed of the wind in the mountain, from the number of cars that circulate at a given moment through a highway to the volume of CO_2 that they are emitting into the atmosphere. Practically any activity of nature and, above all, of human beings, begin to be monitored and quantified. All this immense volume of information, what we know as Big Data, is becoming raw material for the preparation of journalistic information, which is added to the traditional sources of the journalists: personal observation, documents' checking and interviews.

IoT technologies also have a great impact on information processing. One of the most innovative areas is Natural Language Processing (NLP). These technologies, based on the combination of computation, linguistics and artificial intelligence, are giving rise to applications and platforms that allow direct communication between human beings and computers. These technologies are the basis of conversational systems, such as text and/or voice chatbots, and also the origin of applications for automatic writing of texts, known as robots for news writing (Veglis and Maniou 2019). These applications, which have given rise to a new discipline named as 'algorithmic journalism' (Dörr 2016), are expanding technologies in newsrooms, producing unknown challenges for journalists: beyond being simple technologies with a mediating role, they are gaining a new full communicating role, which in the past corresponded exclusively to journalists (Lewis et al. 2019).

The third area where the IoT will impact journalism is on the dissemination of news. The multiplication of mobile devices lived in the last two decades, is giving way to a new ecosystem (Martínez-Costa et al. 2019) where all devices, mobile and desktop, large and small, own and foreign, interact with each other. It is a technological scenario in which the machines are related to each other, constantly exchanging data and presenting to the users the messages that best fit their profile, regardless of whether they are deliberately searched. It is, in short, the rise of a new emerging value in journalism: ubiquity (Pavlik 2014).

2   Technology-Driven News Production and Consumption

Over the last three decades, the digitalization of newsrooms has substantially changed the ways news are produced and consumed. However, the transformation of news production and consumption patterns began long before. At the beginning of the twentieth century, prior to the invention of radio and television, "the newspaper enjoyed the same kind of social monopoly as the railroad did before the coming of the automobile and the airplane [...] It dominated the sphere of information as the train dominated that of transportation" (Smith 1980: 318, as cited in Cunningham and Turnbull 2014).

The role of newspapers in people's social life was crucial and had equivalent importance on the financial strength of news organizations. A century ago, reading newspapers was a widespread social activity, but the emergence of broadcast media soon started to replace that activity, so the growth and circulation of newspapers did not keep the pace of the population. The changes of lifestyle in Western countries seem to have contributed to this steady decline of newspapers, which first symptoms were perceived long before the advent of the interactive networks (Bogart 1972). However, after the impact of broadcast media on news consumption patterns, since the 1990s it has been Internet the main catalyst of the profound changes in these habits.

The multiplication of home computers and derived technologies, such as smartphones, tablets, wearables, and smart speakers, enabled new ways to tell and consume stories. However, they also contribute to cut down the revenue streams of pre-digital media, newspapers in particular. Since embracing web technologies in the 1990s, news organizations are struggling to find their way and adapt their business models to digital media. The quality of information produced by these institutions has a strong bearing on competitiveness and growth, in an age where innovation plays a significant role to engage and interact with the audience (Cunningham and Turnbull 2014). Augmented and virtual reality (VR) is "currently forming the next wave of digitally-driven transformation of economy, culture, and society" (Hassan 2019). Not only them, but there are also great technologies that are driving innovation in newsrooms, such as the use of drones, wearables and smart speakers in journalism, the interactive and visual storytelling formats, the slow journalism, real-time data stories, the news bots, and so on.

2.1 Will IoT Help or Substitute Journalists?

As mentioned above, IoT is already being developed at a fast pace and is expected to dominate the connected devices' landscape shortly. IoT technologies are the next big wave of data-driven innovation (Castillo and Thierer 2015), which may allow tech product companies heading the transformation of what already in the 1970s was named as media or "news ecosystem" (Lozano Bartolozzi 1974). The total number of devices of all kind interconnected will increase exponentially, producing a brand new market for the media, where almost every device, space or surface can—and will—be used as a news platform. The old idea of media as something paper-, radio-, or TV-based will be definitely surpassed, extending it to new 'smart' objects, such as home appliances, wearables, and autonomous vehicles, among others. This pervasiveness of IoT technologies will also transform the way journalists produce news.

Technologies have been key for the news-gathering processes since the late 1800s, first, with the telegraph and later with the telephone. This latter technology became popular in the US newsrooms only by the mid-1930s, affecting the culture of news organizations as it emphasized on mobility and timeliness (Mari 2018). Most of the news workers were reluctant to use the telephone at the beginning. Mainly because it was a complex technology to be adopted by newsrooms and change the skills needed from journalists. The gradual expansion of telephone networks, the standardization of telephone switchboard and the decreasing costs for calls made the telephone an everyday tool in the news organizations.

By the 1920s, the relationship between news workers and technology began to change, giving birth to new long-distance forms of producing news. The technological affordances brought by the telephone, anticipated to some extent by the telegraph, caused internal shifts in the newsrooms' environment, with groups with a higher comfort level, speed of adoption, and ability to innovate taking control over the organizations' workplaces. Therefore, telecommunications—mainly the telephone—can be considered as the first ubiquitous technologies to affect news organizations.

A second technology that increased the ubiquity of media coverage was the automobile. Thanks to the cars, news workers gained autonomy and reach when covering news, while at the same were capable of coordinating better with their peers (Mari 2018).

2.2 More Disruption, Newer Forms of Journalism

Meanwhile, a third newer technology emerged: the mobile, short-range radio, later resulting in a 'radio car', and subsequently the handheld radios. This technology changed the news production in news outlets, which permit editors to keep in constant touch with reporters. The radio car became a reality after World War II, when it got more affordable and commercially available. Thus, several US newspapers began experimenting with that device (Mari 2018).

By 1952, the full automation of newspaper was already envisioned, without manpower of any kind. "The telephone and the radio car, along with some of their ancillary news-gathering technologies, including battery-powered recorders and hand-held radios, both disrupted and strengthened work routines in and out of the newsroom" (Mari 2018: 1385). In other words, sometime in the past "new" technologies already shaped the news, as well as the form they were produced. Thus, there are reasons to believe that this may well happen again, thanks to advanced computing and AI.

Although the use of computers in journalism dates back to the 1950s (Cox 2000), it was only in the late 1980s and early 1990s that the intensive use of databases changed the norms and professional practices of journalism. Back in the 1970s, the father of 'precision journalism', Meyer (1973), already predicted that the use of data would not only help to produce quality content, but also increase readers' engagement. According to Meyer's prediction, the precision journalism, later renamed as computer-assisted reporting (CAR), began to make significant inroads into newsrooms, led by several Pulitzer Prize-winning stories that became a form of professional recognition and validation of practice. That wave finally transformed into what is known today as data journalism (Coddington 2015). Thanks to the advent of the Internet with their large online search engines, as well as to the easier access to databases, data-driven journalism evolved during the 2000s towards its current state. This professional practice represents a democratization of resources, tools, techniques, and methods that in the past were restricted to a few specialists.

Data journalism seems to be a fast-growing trend in the way to future journalism. News organizations need to look into the whole news ecosystem and understand the usefulness of data journalism to adapt news to the new multi-device environment. Data-driven journalism is capable of creating a 'bridge' between technology developers and journalists in newsrooms, expanding their horizons in the working practices and processes, leading them to explore new topics more in-depth (Cairo 2016; Hermida and Young 2017).

Automated journalism, also known as robot journalism (Lemelshtrich 2018), is another emergent practice within an established field of practices that is journalism. "An example of automated journalism in which a program turns data into a news narrative, made possible with limited—or even zero—human input" (Carlson 2015: 416). These technologies are being disrupted by companies that are not media organizations, such as Applied Semantics, Automated Insights, and Narrative Science (Caswell and Dörr 2017).

When in March 2014 *Los Angeles Times* reported a 4.7 magnitude earthquake in the city three minutes after the rumbling stopped, no one could imagine the news story was written by a robot. However, the information that had been reported by nobody, written by nobody and published by nobody, was the news everyone was reading (Salaverría 2017: 16). Since then, many news organizations around the world have embraced automated news writing technologies. The system behind these technologies is the natural language generation (NLG), which involves the automatic creation of text from digital structured data, a technology that has been developed and commercialized over the past decade (Carlson 2015). Today, the most obvious examples of automated journalism are in routine sports and financial news. Several reasons explain why robotic journalism has not expanded yet to other topics and to even more complex journalistic writing. One of these reasons is the absence of appropriate data in those topics, which so far has put up a barrier in the development of this technology. This limitation might be soon overcome with the advent of IoT and AI to the news industry (Caswell and Dörr 2017).

Automated journalism's ability to produce news stories without human mediation raises questions about the future of journalistic labor (Carlson 2015; Caswell and Dörr 2017). These concerns come from the sensitivity of the ongoing workforce cuts and layoffs in the media industry. These fears are not new; they date back to the earliest days of nineteenth-century industrialization, when some people's jobs began to be replaced by machines. Today's situation is a step forward in that same old path: "The long-running trope of human versus machine is now complicated by developments in artificial intelligence regarding the mimicking of human storytelling" (Carlson 2015: 424).

One of the leading companies in automated writing technologies, Narrative Science, has indicated how its technology will alter journalism. This American company highlights that the tool will not replace journalists, but it will free them up to pursue other stories. It claims a new approach to news production, as it enables novel ways to collect, distribute, and consume news. According to this company, technology is saving 'boring' activities to humans, so that they can focus on more relevant activities. It is no less true, however, that this dehumanization of news production process raises questions about who and how will control the news flows in the future.

Considered from the optimistic side, the renewal of news production forms is an opportunity to scale and personalize news stories. Before the automation, the limited availability of staff and space in the newspapers caused publishers "to base coverage decisions on ideas of newsworthiness to attract its desired audience both maximally and efficiently" (Carlson 2015: 425). Thus, today's ongoing robotic journalism has the potential to drastically alter the conditions of news production and consumption. It may reduce the production costs of a large portion of news reporting, allowing a constant update of information without current difficulties. It may adapt automatically the form of this news reporting, based on their response rate by the audience. Finally, the future of automated news opens the possibility to create multiple customized versions of the same story for individual users. Even though all these changes might result in many more stories published but with fewer hits each, the overall web traffic will surely increase. That model is impossible today due to human labor costs. However, besides the impact on labor, there are also other concerns: as the automatization expands, so will do also the threats of echo-chambers, which limit the diversity of opinions people are exposed to (Cardenal et al. 2019).

Computational thinking, an ability to think abstractly about the use of language in parallel of its logical rules and practical language skills (Wing 2006), will be a necessary skill for journalists working with automated news production. "Journalists should be able to work with structured events, narratives, and narrative abstractions without any coding skills, database skills, mathematical skills, or other technical expertise" (Caswell and Dörr 2017: 492). Off-the-shelf solutions will help journalists who don't have the skills to develop automated stories without coding skills, mathematics skills or any other computational expertise. However, learning to code is essential for journalists to get computational thinking and being able to communicate and work effectively with other actors involved in news production (Caswell and Dörr 2017).

Furthermore, the evolution of technology is allowing publishers to create new forms of multimedia news storytelling. There are already autonomous video production systems, which automatically combine and edit texts, pictures, and short videos through out-of-the-box solutions. "Increased automation is probably essential to fulfilling journalism's societal mission and will therefore probably become a more common aspect of the practice of journalism" (Caswell and Dörr 2017: 493).

The technological affordances provided by IoT and AI are expanding thanks to new devices, such as drones, wearables and smart speakers. These objects will be not only tools to access information, but also to provide very detailed audience data to publishers and advertisers. In brief, technology can alter power relationships between actors involved in the development of new technologies and innovation in newsrooms.

The underlying drivers of the technology revolution are defined by an increase in "processing power, storage capacity, and networking capabilities", along with "the digitalization of data and assembly of 'big data' repositories" (Castillo and Thierer 2015: 2). As the telephones or radio-cars took some time to be incorporated by media companies, the same applies to IoT technologies (Mari 2018). These will happen neither overnight or smoothly, but the early adoption by some newsrooms will surely make others to catch the pace and strengthen their will to embrace these communication technologies.

As pointed by Belair-Gagnon et al. (2017: 1235), "historically, innovation in news production has often occurred outside traditional newsrooms' settings and, in many cases, led to the creation of new forms of journalism", such as photojournalism and the rise of photojournalism magazines. Today we can observe again that same process: unmanned aerial vehicles (UAV) or unmanned aircraft systems (UAS), more commonly known as drones, have become an important case of disruptive journalistic technology in news organizations. These devices were developed outside newsrooms, but made their way into established institutions in journalism.

Camera drones first appeared in the news media in 2011, during the riots in Warsaw, Poland, and thereafter in the event of the Occupy movement (Lauk et al. 2016; Uskali 2018). Since then, drones have shown potential to provide productivity gains and cost savings to data gathering in journalism. UAVs extend the surveillance capabilities of media, giving access and providing tools for new forms of storytelling.

Furthermore, the possibility to have better and more precise visualization without human risk has led to the gradual incursion of drones in news work. The range of perspectives and aerial element provided by images produced by UAVs develop new forms of witnessing news and access remote areas, or places that were inaccessible by journalists before. The technology has progressed to implement autonomous algorithmic control of drones, which offers a cost-efficient way to capture images and data through sensor-based UAV journalism.

2.3 Journalism Every Time, Everywhere and in Every Form

Since the beginning of the Internet age, information has become an easy to find commodity. Today's new technologies promise a more real-time, multi-platform news consumption, accessible to almost everyone and anywhere via Internet live-streaming (Uskali 2018). This type of panoptic broadcasting is already present in social networks.

Over the last few years, social media networks, one of the main expressions of participatory culture (Jenkins et al. 2013), have had a deep influence on the birth of ubiquitous journalism, bringing new ways of disseminating the news. For instance, Twitter has become a common tool for journalists around the world to cover breaking news events. Facebook promotes algorithmic news feeds that suggest personalized content, according to users' browsing history. As indicated by Uskali, "ubiquitous journalism favors the social media platforms that can offer the largest audience, and right now that is clearly Facebook" (Uskali 2018: 243).

New technological affordances have caused shifts in the ways people consume news, creating new ways to reach audiences. New devices like smart speakers and smartwatches are some of the innovations brought to the news industry, which might be a game changer, especially in live news situations. "The smart speakers offer the potential to challenge the foundation of radio, turning broadcasts into conversations, changing the stories people hear and reading individualized streams of information" (Bullard 2019). Smart speakers can replace or augment the functions of a radio or phone, powering them with AI. Through voice commands, users can consult their devices not only about the news but also to play music, find recipes or answer simple questions. Therefore, voice is pushing the audience's connected lives to new habits and leading users to rethink how they interact with technology. The promising future of smart speakers might influence the whole news industry, transforming voice in a more dominant interface (Newman 2018).

In the US, in 2018 there was still only 18% of smart speakers' owners listening every day to the news and 22% were using these devices to listen to podcasts (Newman 2018). There is a great opportunity for a news organization to boost this technology, which is still in its infancy. Owners of smart speakers complain about "hearing the same story told by different outlets in different ways, possibly at different volumes"

(Bullard 2019). So far, the lack of local content and limited personalization does not allow to expand audience's engagement. To overcome this, publishers need to design news updates that take advantage of smart speakers' unique assets. The challenge is to find out what users expect to hear and how it should sound. The possibility of having a conversation with the listener makes this device more genuine, creating an interactive environment that allows engaging further audiences. The future might be answering to topical news questions and provide stories targeted to specific listeners. With the data provided by searching terms from millions of users flowing in, news organizations could cater their work to make the content more relevant. Thus, the "most successful ways of telling stories in an algorithmically curated, the voice-based news feed will be determined by user data" (Bullard 2019).

Chances are that smart speakers will lead to novel standards in the news production to capture the diverse range of news, views, and opinions from a wide range of sources, so that, journalism can develop the critical thinking of the audience. For this reason, human curation should be a big part of what algorithm is doing to ensure that the stories aren't subject to personalization or algorithmic bias. The next step in the growing evolution of audio news reporting might be the completely synthetic voices, which news organizations will be using text-to-voice technology to produce their news bulletins. The relationship between humans and machines will have become even more valuable to guarantee that the algorithms are performing correctly and distributing stories properly.

Wearables are a subset of technologies that integrate networked devices into portable accessories. These applications can be found in watches, clothes, glasses, just to name a few, and they are the fastest-growing segment of IoT. Even the possibility of having under-skin devices is envisioned. This wide array of technologies promises to connect 'smart devices' with massive processing power and speedy.

So far, smartwatches are the most popular wearables, allowing a constant and ubiquitous connectivity. Apple and Samsung, with their respectively Apple Watch and Samsung Gear, are leading this disruption. Other examples of less advanced examples are the Jawbone and Fitbit, with allow individuals to measure and share their daily fitness activity (Castillo and Thierer 2015). These devices are not only a means to be informed but also data producers, which offer an interesting tool of information gathering. Wearables lead to an interconnected platform based on a dominant logic of constant news updating through notifications, which enhances the capacity to reach users directly. Informative alerts were first implemented by mobile devices through SMS or MMS. Smartwatches seem to be much more effective in breaking news situations and also in delivering financial and sports news (Uskali 2018).

The data generated by audiences are also a source of real-time information, which could be used for media organizations. To some extent, users could become into involuntary reporters, providing data about events they are part of or, for instance, submitting data about meteorological conditions (Castillo and Thierer 2015; Silva-Rodríguez et al. 2017).

This post-mobile wave of technologies also creates new ways to engage and generate empathy with the audience. The immersive journalism is a result of the novel capabilities made possible through the deep progress in the fields of video. These technological innovations got extremely popular in 2016 with virtual reality (VR) and augmented reality (AR) (Nordrum 2016). The immersive journalism is a form of news storytelling where the audience can gain a first-person experience of events (Mabrook and Singer 2019).

Currently, the immersive experience can happen in two different ways. The first one is 'virtual reality', a term coined in the 1980s by Jaron Lanier, a musicologist and engineer. VR is awaiting to form the next wave of digitally-driven transformation of the economy, culture, and society. This technology creates immersive atmospheres that can be alike to or radically different from the real world (Terdiman 2018), which have as central elements the immersion and interactivity. 360-degree video, also known as immersive video, "enables users to look in every direction, thus placing them 'inside' an environment" (Mabrook and Singer 2019). Usually shot by using an omnidirectional camera or a collection of cameras, these videos can be engaging and even emotionally impactful, but they are essentially just a new form of filmmaking (Hassan 2019; Mabrook and Singer 2019).

The second type of immersive journalism is brought by augmented reality. AR is an evolution built on the "transformation that photography as a media of storytelling brought to journalism in the 19th century" (Hassan 2019: 2). Overlapping the virtual world with the real world, the AR promises to capture attention and empathy conducting through a semi-real world, the viewer 'augments' the space from physical reality. "Duty of the journalist is to bring the far to the near" (Hassan 2019: 15). Thus, technology has this function and helps journalism to bring a 'reality' or a 'story' that people wouldn't have access to it.

Beyond VR and AR, there are other emergent technologies that promise immersive experiences. One of them is the autonomous car, which will be a networked vehicle connected to wireless communication and dynamic programming. This will turn drivers into passengers, freeing them up to do other tasks. Just as the appearance of the car as a means of transportation changed the habits of citizens and served to popularize the radio, these smart cars might change again the way we consume news. The possibility of being transported by a self-driven car may produce significant shifts in the way current drivers spend their time, allowing them to do other things, such as consuming news with no need of paying attention to the traffic. The same might happen to automated drones, which might employ similar networked concepts to automate aerial operations. Thus, news organizations may increase their productivity and gain cost savings to gather data and produce news stories (Castillo and Thierer 2015).

It's not yet clear how these technologies will cohabit with each other. But there's no doubt about the disruption of novel technologies will create new possibilities and challenges for journalism.

3 The Advent of Ubiquitous Journalism

All these technological transformations are driving journalism towards a new ubiquitous paradigm. This is an emerging form of journalism, characterized by an expansion in the modes of production and consumption of information, thanks to the intensive use of algorithmic systems and the personalized distribution of content, in all types of digital devices. These technologies are taking journalism to another dimension: first came media digitalization, then participatory journalism and now, finally, comes news robotization. It is a model of journalism that transcends the boundaries of the classic news media outlets—press, radio, television, Internet—to convert, at least potentially, any digital device into a media platform.

According to what it has been explained in the previous pages, *ubiquitous journalism* can be defined as the *type of journalism that, thanks to the intensive use of algorithmic systems and artificial intelligence, disseminates news in multiple digital devices produced by journalists, users and robots, so that it is consumed anywhere and anytime by the public, through a constant flow of personalized and multisensory information.* As this definition points out, ubiquitous journalism is characterized by five main features:

(1) expanded news production, with content elaborated by three types of sources: journalists, users, and robots;
(2) multi-device access to information, through every kind of visual, sound, and haptic (Parisi et al. 2017) interfaces;
(3) constant flow of information, without restrictions of time or space;
(4) personalized distribution of information through artificial intelligence systems, capable of targeting specific content to each user, based on their personal preferences and digital history;
(5) immersive news storytelling, with expanded multimedia content targeted to multiple corporal senses.

These five elements have been explored separately in recent years by some media companies, usually the most advanced. The great novelty now is that the five elements can be combined all together, multiplying their impact and opening unknown possibilities for the news media.

These developments are full of opportunities, but also threats, for journalism. Continuing the path that the media industry undertook in the 1990s, when it launched the first web publications, now it can take a step forward in the exploration of innovative ways of presenting news and interacting with the public. Locative technologies (Goggin et al. 2015) and the processing of large volumes of data through algorithmic systems will open spectacular possibilities to the media for the dissemination of personalized news, aimed at any physical environment, device or user profile.

These are, without a doubt, opportunities that journalism must explore. However, at the same time, these technologies are an undeniable menace to people's privacy and even to their right to be fairly informed, as these technologies may distribute news contents according to hidden commercial and/or ideological interests. Transferring from journalists to machines not only the capacity to produce news, but also to rank and distribute them among specific audiences is a phenomenon in which long-term impact we are far to envision yet. The news agenda, more and more personalized, will be largely determined by the robots. The liquid element called information that surround us and where we live may no longer be controlled by journalists.

Acknowledgements This chapter has been funded by two research projects: JOLT—Harnessing Data and Technology for Journalism (H2020—MSCA-ITN-2017; grant number: 765140), and DIGINATIVEMEDIA - Digital Native News Media in Spain: Characterisation and Trends (Spanish Ministry of Science, Innovation and Universities; grant number: RTI2018-093346-B-C31).

References

Bauman Z (2000) Liquid modernity. Polity Press, Cambridge

Belair-Gagnon V, Owen T, Holton AE (2017) Unmanned aerial vehicles and journalistic disruption. Digit J 5(10):1226–1239

Bogart L (1972) The age of television: a study of viewing habits and the impact of television on American life. Ungar Publishing Company, New York

Bullard G (2019) Smart speaker use is growing. Will news grow with it? Nieman reports. Retrieved 13 May 2019, from https://niemanreports.org/articles/reimagining-audio-news/

Cairo A (2016) The functional art. New Riders Publishing, Berkeley

Cardenal AS, Aguilar-Paredes C, Cristancho C, Majó-Vázquez S (2019) Echo-chambers in online news consumption: evidence from survey and navigation data in Spain. Eur J Commun (Online first, 23 Apr)

Carlson M (2015) The robotic reporter: automated journalism and the redefinition of labor, compositional forms, and journalistic authority. Digit J 3(3):416–431

Castillo A, Thierer AD (2015) Projecting the growth and economic impact of the internet of things. Mercatus Center

Caswell D, Dörr K (2017) Automated journalism 2.0: event-driven narratives. J Pract 12(4):477–496

Coddington M (2015) Clarifying journalism's quantitative turn. Digit J 3(3):331–348

Cox M (2000) The development of computer-assisted reporting. Newspaper division. Assoc Educ J Mass Commun SE Colloquium 2030(305):22

Cunningham SD, Turnbull S (2014) The media and communications in Australia, 4th edn. Allen & Unwin, Sydney

Deuze M (2008) The changing context of news work: Liquid journalism for a monitorial citizenry. Int J Commun 2(18):848–865

Dörr KN (2016) Mapping the field of algorithmic journalism. Digit J 5(8):1044–1059

Goggin G, Martin F, Dwyer T (2015) Locative news: mobile media, place informatics, and digital news. J Stud 16(1):41–59

Gossage HL (1967) You can see why the mighty would be curious. In: Stearn GE (ed) McLuhan, hot & cool; a primer for the understanding of & a critical symposium with a rebuttal by McLuhan. The Dial Press, New York, pp 15–34

Greengard S (2015) The internet of things. MIT Press, Cambridge

Hassan R (2019) Digitality, virtual reality and the 'Empathy Machine'. Digit J (Online first, 2 Jan)

Hermida A, Young ML (2017) Finding the data unicorn. Digit J 5(2):159–176

Jenkins H, Ford S, Green J (2013) Spreadable media: creating value and meaning in a networked culture. New York University Press, New York

Lauk E, Uskali T, Kuutti H, Hirvinen H (2016) Drone journalism: the newest global test of press freedom. In: Carlson U (ed) Freedom of expression and media in transition: studies and reflections in the digital age. Nordicom, Gothemburg, pp 117–125

Lemelshtrich LN (2018) Robot journalism: can human journalism survive?. World Scientific Publishing, New Jersey

Lewis SC, Guzman AL, Schmidt TR (2019) Automation, journalism, and human–machine communication: rethinking roles and relationships of humans and machines in news. Digit J (Online first, 23 Apr)

Lozano Bartolozzi P (1974) El ecosistema informativo. Introducción al estudio de las noticias internacionales. EUNSA, Pamplona

Meyer P (1973) Precision journalism: a reporter's introduction to social science methods. Indiana University Press, Bloomington

Mabrook R, Singer JB (2019) Virtual reality, 360° video, and journalism studies: conceptual approaches to immersive technologies. J Stud (Online first, 17 Jan)

Mari W (2018) Technology in the newsroom. J Stud 19(9):1366–1389

Martínez-Costa MP, Salaverría R, Breiner J (2019) El ecosistema que viene. In: López García X, Toural-Bran C (eds) Ecosistema de cibermedios en España. Tipologías, iniciativas, tendencias narrativas y desafíos. Comunicación Social, Salamanca, pp 225–240

Newman N (2017) Journalism, media, and technology trends and predictions 2017. Oxford Reuters Institute for the Study of Journalism

Newman N (2018) The future of voice and the implications for news. Oxford Reuters Institute for the Study of Journalism

Nordrum A (2016) The fuzzy future of virtual reality and augmented reality. IEEE Spectrum, New York

Parisi D, Paterson M, Archer JE (2017) Haptic media studies. New Media Soc 19(10):1513–1522

Pavlik JV (2014) Ubiquidade: O 7º princípio do jornalismo na era digital. In: Canavilhas J (Org) Webjornalismo: 7 caraterísticas que marcam a diferença. Livros LabCom, Covilhã, pp 159–183

Pew Research Center (2014) The internet of things will thrive by 2025. Retrieved from http://www.pewinternet.org/2014/05/14/internet-of-things/

Salaverría R, Sádaba C (2003) Towards new media paradigms: content, producers, organisations and audiences. Ediciones Eunate, Pamplona

Salaverría R (2017) Allá donde estés, habrá noticias. Cuad de Periodistas 35:15–22

Silva-Rodríguez A, López-García X, Toural-Bran C (2017) iWatch: the intense flow of microformats of "glance journalism" that feed six of the main online media. Rev Lat de Comunicación Soc 72:186–196

Shenk D (1997) Data smog. Surviving the info glut. Technol Rev 100(4)

Terdiman D (2018) Why 2018 will be the year of VR 2.0. Retrieved from https://www.fastcompany.com/40503648/why-2018-will-be-the-year-of-vr-2-0

Uskali T (2018) Towards journalism everywhere. In: Daubs MS, Manzerolle VR (eds) Mobile and ubiquitous media: critical and international perspectives. Peter Lang, New York, pp 237–247

Veglis A, Maniou TA (2019) Chatbots on the rise: a new narrative in journalism. Stud Media Commun 7(1):1–6

Wing JM (2006) Computational thinking. Commun ACM 49(3):33

Ramón Salaverría Ph.D. and associate dean of research at the School of Communication, University of Navarra (Pamplona, Spain), where he heads the Digital News Media Research Group. Being author of over 200 scholar publications, his research focuses on digital journalism and media convergence, both in national and international comparative studies. His most recent book is *Ciberperiodismo en Iberoamérica* [Digital journalism in Ibero-America] (2016), a comprehensive analysis of the evolution of digital media in 22 Latin America countries, Spain, and Portugal.

Mathias-Felipe de-Lima-Santos Ph.D. Candidate on Branding Data Journalism at the University of Navarra under JOLT project, a Marie-Skłodowska-Curie European Training Network. In early 2018, he was visiting researcher at DMRC of Queensland University of Technology (QUT). He completed the EU-funded master's qualification Digital Communication Leadership offered by University of Salzburg and Aalborg University. His research interests lie in the ongoing changes to journalistic practice, with a particular focus on business models, digitalization, big data and data visualization.

From Data Journalism to Robotic Journalism: The Automation of News Processing

José Miguel Túñez-López, Carlos Toural-Bran
and Ana Gabriela Frazão-Nogueira

Abstract The 21st century is a reaffirmation of automation and through big data and data journalism begins to speak of the 'robot journalism', the 'automated journalism' and the weight of 'cognitive journalism'. This article refers to how Artificial Intelligence (AI) is beginning to occupy a field traditionally dominated by the human factor in the management of information relations between organizations, the media and society by the application of data mining to generate algorithms that make it possible to automate the management and derive it to the work of bots in the elaboration of news. It also addresses how the journalistic profession lives apparently oblivious to the robotization of the newsrooms, although the origin of mass automation dates back to 2014 when Associated Press with Automated Insights and Zacks Investment Research generated 3000 news about 'corporate profits'.

Keywords Algorithm · Robot journalism · Automated journalism · Artificial intelligence

1 From the Algorithm to Analyze to the Algorithm to Write

The news written automatically could win the Pulitzer in a few years. The irony of Lindén (2017) and Levy (2012) is also a prediction of the growing impact of the algorithmic generation of informative content that is transferred to the audience as content of the media agendas. The interest for the contents elaborated by machines is a recurrent theme in the present, but in reality it corresponds with a way of producing news that began at least half a century ago with meteorological information, included in the weather section (Meehan 1977; Glahn 1970) and, already in the final years of

J. M. Túñez-López (✉) · C. Toural-Bran
Universidade de Santiago de Compostela, Santiago de Compostela, Spain
e-mail: miguel.tunez@usc.es

C. Toural-Bran
e-mail: carlos.toural@usc.es

A. G. Frazão-Nogueira
Universidade Fernando Pessoa, Porto, Portugal
e-mail: ana@ufp.edu.pt

© Springer Nature Switzerland AG 2020
J. Vázquez-Herrero et al. (eds.), *Journalistic Metamorphosis*,
Studies in Big Data 70, https://doi.org/10.1007/978-3-030-36315-4_2

the 20th century, in some of the topics that achieved space in the pages of economics or sports (Meehan 1977).

Is then in the last decade of the 20th century that could be located the beginning of the change of trend or the start of the real takeoff of the automation of news, with the resource to generate financial software, data, and news content that some companies, like Bloomerang LP, start to offer to a portfolio of clients that includes some news agencies and media as representative as Thomson Reuters or *The New York Financial Press* (Winkler 2014).

By following the trail of the irruption of computers in the creation of informative content, the path traced goes beyond the computerization of the newsrooms. Some authors link the origin of the automated generation of news to data journalism and, as in the case of Gynnild (2014), even point Philip Meyer as a pioneer.

The starting point is in the computer-assisted reports (CAR) that are identified as the launch for what later would be known as 'precision journalism', defined by Meyer himself as the application of social and behavior research methods to the exercise of journalism through a deep exploration of databases, surveys, and a general combination of informatics and social sciences (Meyer 1975).

Already in the first years of the 21st century two initiatives are identified that could be considered pioneers in the visualization of data as news. On the one hand, the Chicago Crime—Google map mashup, launched in 2005, and, on the other hand, crime information in real time by *Los Angeles Times*. Both represent a step forward that turns the visualization into news and that is reinforced when this same newspaper, *Los Angeles Times*, resorts shortly after, in 2007, to Quakebot, an algorithm that uses data from the US Geological Survey to prepare information from a previous template.

This experience goes one step further and not only computerizes the news, but also automates that the news written by the robot is already published directly, if the earthquake is less than magnitude 6. There is a coincidence, however, in considering that the origin of mass automation is initiated by Associated Press with Automated Insights and Zacks Investment Research, in 2014, to generate 3000 news on 'corporate profits'.

In this synthesis of changes there is an important transformation. It is not just about the process' informatization, but the scope of that computerization to the phases of the process that had not been affected before. It is the step of the algorithm to analyze to the algorithm to write in a constant evolution of change that, in this case, has been favored by the progression of Web 2.0 to 3.0, characterized by the appearance and consolidation of the web semantic and by the application of Artificial Intelligence for the storage and processing of data that facilitate the passage from data journalism to computer journalism. "Computational journalism works primarily through the abstraction of information to produce computable models, while data journalism works primarily through the analysis of data together to produce data-oriented stories" (Stavelin 2014).

2   Journalism and Big Data

Carlson (2014, referencing Mayer-Schönberger and Cukier 2013) influences this differentiation by ensuring that the automated generation of news comes to be the result of the intersection between journalism and big data. In his opinion, computers can be used for information retrieval and data mining processes can be used to discover new knowledge of structured and unstructured random data silos (Wölker and Powell 2018) and allows, in addition, to complete the process, introducing interactivity with consumers (Flew et al. 2012).

The robotization of newsmaking has been a constant since the computerization of the newsrooms began. The change of the typewriters by computer and the replacement of news reception systems by journalist computer reception systems were celebrated as a technological advancement that improved the process of construction of the agenda but opened the way for the redefinition of professional profiles that participate in the news production process, both in the planning, connection, layout, printing and product distribution parts.

The change has been progressive and continuous, but it has been done in a way that always seemed to affect the mode of information production but not the production of content directly and, even today, journalists do not have a clear perception that they already share time or space of informative emission of the contents that they elaborate and those that are obtained from automatic way through algorithms.

A recent study based on interviews with 366 Spanish journalists confirms that among professionals of journalism there is still no clear awareness that the generation of news through algorithms has ceased to be a possibility to be a reality and is "even unknown that some international media and agencies have already replaced their editors with computer applications to produce content that they transmit to their audiences" (Túñez-López et al. 2018: 756).

Evidently, the 20th century has marked an increase in the speed of transformation as its decades advanced. At the turn of the 21st century one can already speak of a reaffirmation of automation and, as seen, through big data and data journalism, it was possible to start talking about the robot journalism, automated journalism or cognitive journalism.

3   New Productive Routines

Artificial Intelligence began by changing the routines of the journalist by automating some of the functions of search, classification or processing of information and has begun to install itself already in its functions by also covering the tasks of news writing. The use of bots to generate text "is the pinnacle of a process of decades of automation in newsrooms" (Lindén 2017) that began at the end of the "80s of the 20th century" (Túñez-López et al. 2018: 751).

The result is what has been labeled as automated journalism, which has been defined as "algorithmic processes that convert data into informative narrative texts with limited or no human intervention beyond the initial program" (Carlson 2014: 417) or as the "process of using software or algorithms to automatically generate news without human intervention, after the initial programming of the algorithm" (Graefe 2016) interrelating "the fields of informatics, social sciences and communications" (Flew et al. 2012: 157).

Other authors have chosen to refer to it as algorithmic journalism (Dörr 2016) and robot journalism (Oremus 2015), but always identifying it as a technological solution to produce news or other journalistic tasks such as reports, therapeutic or even analysis and visualization of data (Carlson 2014; Gao et al. 2014; Young and Hermida 2015; Shearer et al. 2014; Broussard 2015).

Independently of its denomination, the elaboration of news through computer programs that give autonomy of data interpretation and writing of texts to the computers can be carried out through the identification of repeated writing routines that can be identified and codified because it is based on the simulation of natural language through software that allows the robotic creation of informative texts elaborated by computer, but with identical characteristics to another elaborated by a human.

The creation of algorithms and programs that allow writing with increasingly autonomous machines has been a constant in technological development because, as pointed out by Rivera-Estrada and Sánchez-Salazar (2016), for humans to endow them with autonomy has represented a dream that the Artificial Intelligence is performing in an increasingly visible way since its application in online environments affect the daily life of citizens to the extent that they produce and help to shape their referent of reality through the news that receive.

4   Bots and Algorithms in the Newsrooms

In a simple definition, Artificial Intelligence is oriented to emulate (imitate or repeat) intelligent behavior, but it does so through computational processes (Córdoba-Guardado 2007). Its irruption in the environments of mass media content, or in a more generic way, in communication, coincides with the rise of a new environment, the result of universalization and massive access to the internet, which multiplies the possibility of obtaining and disseminating data that, at the same time, can be treated massively through computer systems.

The internet has forced a restructuring of the media, has changed its methods of financing itself and producing content, has transformed the way of relating to audiences and has allowed new platforms to appear, the online media, in which one works with hypertextuality, interactivity, and multimedia. That is, it has upset the profile of journalism and expanded the operative functions of journalism. Parallel to the changes in products and support, the current phase is characterized by the technological development that goes into the creation of contents and the writing of news based on algorithms to be generated by computer.

Expressed in a metaphorical way, this step forward converts the algorithms into the new journalists. Defined as a finite series of specific descriptive norms, algorithms are the step-by-step abstraction of a procedure that takes an input and produces a result to achieve a defined product (Diakopoulos 2014).

Applied to journalism or to the generation of informative texts to disseminate in any medium and not only in the media, Anderson (2011) and Carlson (2014) explain how algorithmic formulations can prioritize, classify and filter information and even be applied as metrics of audience's analysis to determine topics to cover and, according to the information obtained or provided in databases, to write stories.

The use of algorithms allows, then, machines to become programmable and autonomous generators of textual journalistic products, graphic or infographics, from data. As Graefe (2016: 5) points out:

> once the algorithm is developed, it automates every step of the news production process, from the collection and analysis of data, to the creation and publication of these. [...] In this context, algorithms can create large-scale content, personalizing it to the needs of an individual reader, faster, cheaper and potentially with fewer errors than any human journalist.

One of the important differences is that automated journalism does not work directly on the reality defined by facts, but on a reality encoded mainly in data on which the algorithms act. It is an important nuance because it allows deriving interest to four aspects:

(i) the ability of the AI to replace the cognitive part of the journalistic work and code it algorithmically;
(ii) the process of preparing the databases;
(iii) the rules of construction of the algorithm;
(iv) the robots involvement in the possible generation of false stories.

Attention and social debate are surely centered on the current perverse use of automation to appropriate all the symbols of news and their codes and channels of dissemination to introduce into the public sphere fake information. Without entering into the debate that the fake news can not be considered news, because the news reflects a true story without previous intentionality or conscious of cheating, this analysis focuses on the computerization of the production of real news. That is, tasks performed by machines to be incorporated into the informative story as a part of the narrative of current references, transmitted by the media.

The use of computers, bots and/or algorithms to produce news content is an alternation of the newsmaking process that opens a new path of attention because it is no longer just a matter of debating the convenience or suitability of replacing the individual with the machine and the possibility to automate the ability to analyze, interpret and narrate. The robotization of newsmaking generates the story about data not about the fact itself and this forces to guide towards the creation of new spheres of control over the information that is published.

In fact, voices have already been raised that consider it necessary to thoroughly review the ethical, moral and operative considerations of computer-generated news because Artificial Intelligence tends to concentrate more power in the hands of those who are already powerful "as we have already seen in Google, Facebook and Twitter" (Lindén 2017: 73).

The replacement of journalist by robots to generate news is a hot topic in academic research, especially since the beginning of this decade from the work of Powers (2012) and Karlsen and Stavelin (2013) on the impact in journalism of advances in technology, or the contributions of Flew et al. (2012), on the use of computers as tools to increase interactivity with consumers.

There are relevant publications on the application of Artificial Intelligence to the elaboration of news such as Kim et al. (2007), Matsumoto et al. (2007), Van Dalen (2012), Clerwall (2014), Edge (2014), Karlsen and Stavelin (2014), Latar (2014), Napoli (2012), Stavelin (2014), Carlson (2014), Oremus (2015), Lecompte (2015), Dörr (2016), Graefe (2016), Fanta (2017), Hansen et al (2017), Lindén (2017), Marconi and Siegman (2017), Usher (2017), Salazar (2018) and, among others, Wölker and Powell (2018) reflecting a growing interest in scientific research for the robotic development of news stories.

5   Investigations on AI and Journalism

For a rigorous study of the informative automation it is necessary to stand out, among them, the investigations that, at the time, were novel for providing concrete cases of robotization. Thus, Graefe (2016) and Dörr (2016) identify media in which automated news were already being used, the Fanta (2017) report refers to the use of computer generated news in the European news agencies and Renó and Renó (2017) contribute on the use of algorithms to generate stories in media and agencies.

The contributions of Clerwall (2014), which analyzed the differences in the perceived quality of 46 Swedish students in two versions of an article about a game of American football with human and robotized authorship, establishing a comparison whose results demonstrated that, for the public, there were no important differences between both texts. Graefe et al. (2016) take up the case to try to answer the reasons for the results obtained.

Reports such as the *Digital News Report* 2017 of the Reuters Institute and the University of Oxford reinforce the idea that there are no clear preferences of the public among the contents elaborated by machines or by humans, and they even opt slightly towards the news selected by algorithms. According to the data of this report, in general, 54% opted for automated selection compared to 44% opting for the one made by humans. When the data are revised by age, among those under 35 years of age, the preference for the information proposal made by a robot increased to 64%.

Other studies that focus on issues such as opportunities and challenges of journalism on accountability centered on algorithms (Diakopoulos 2014); the public's perception of informational texts produced through Artificial Intelligence (Graefe 2016); on the benefits of personalizing local information thanks to structured data (Lecompte 2015); on the response of the media to the automation of content (Lindén 2017) and even research in the educational field such as the analysis of Slater and

Rouner (2002) on the response of groups of people of different levels and ages to texts made by journalists and robots.

Other investigations are oriented to know what is the perception that professionals have of the information of algorithmic inspiration in the newsrooms. Van Dalen (2012) analyzed the reactions of journalists to the launch of StatSheet, a network of sports websites written by machines; Carlson (2014) examined how journalists wrote about the text generation software published by Narrative Science; Young and Hermida (2015) examined the appearance of news about computer crimes in *Los Angeles Times* and Thurman et al. (2017), interviewing ten journalists from media such as *CNN*, *BBC* or Thomson Reuters to get their impressions on various articles that had been generated in an automated way. Túñez-López et al. (2018) studied the degree of knowledge of the penetration of robotization and the attitude of Spanish journalists.

The results are convergent in their description, although the computerization of the generation of news generates conflicting reactions. Critics with the use of bots suggest that algorithmic journalism could represent the "most unsettling model, both for communication and for democracy" (Anderson 2011: 541), a challenge to the authority of traditional journalists (Usher 2017).

Those who question the robotization of newsrooms argue that the use of algorithms to create news is a disruption with the idea of what is journalism not only because the bots can not ask questions, determine causality or form opinions, but because they may be inadequate to fulfill the function of 'guardian dog' (Strömbäck 2005) since it is not possible to think of algorithms that become "guardians of democracy and human rights" (Latar 2015: 79). They also say that robotization will have a negative impact on employment, because it will mean the elimination of jobs, and in the content, since it can mean that the media pass to emit or publish insipid and repetitive news.

However, in general, most studies agree that, as Carlson (2014: 418) points out, journalists react to technological innovation "in a complex way, from fear (…) to reinvention". The most optimistic argue that, with the algorithms, the content will be more attractive and the news written by computer "could potentially increase the quality and objectivity of the news coverage" (Graefe et al. 2016: 597) or defend that the automation allows the content to be produced faster, in multiple languages, in greater numbers and possibly with fewer errors and biases.

Clerwall (2014) adds that robotization is perceived as a form of collaboration with the human journalist or a distribution of the workload since it frees that professional from tasks. In this same direction, Flew et al. (2012) explains that when the machine frees the journalist from the work of obtaining the data, allows him to focus on the verification of news, on counteracting 'false news' (Graefe et al. 2016) or on making exhaustive or investigative reports while routine tasks are covered with algorithms.

Economic reasons are also indicated because they would allow the media to offer a wide range of stories at a minimal cost (Van Dalen 2012) and reasons of opportunity: "if journalism must be cyborg, robots must be part of the panoply of professionals; machines that help them increase the power and scope of their journalism, and not rivals that endanger their jobs" (Cervera 2017: 108–109).

Other research works, such as those by Chu et al. (2010), Tavares and Faisal (2013), Dickerson et al. (2014) or Ferrara et al. (2016), have been oriented to review the use of robots in social networks, especially, in the extraction of characteristics such as temporal activity, network structure and user sentiment to develop automatic learning classifiers that allow detecting the robotic management of the profiles. Keeney (2015) focuses on analyzing how hyper-targeting social network users can track their fingerprints to suit their preferences.

Defined as 'automated social actors' the bots in networks are oriented to simulate human behavior (Lokot and Diakopoulos 2016) in the management of content and interactions (Hwang et al. 2012: 40) and to spread positive content or expand fakes and generate unwanted relationships (spam). More than generating news, they are therefore oriented to participate in the dissemination on social platforms (Lokot and Diakopoulos 2016), to retransmit or to add web content (Mittal and Kumaraguru 2014; Starbird et al. 2010) and to identify events of journalistic interest for its later diffusion (Steiner 2014) since the algorithms can be adjusted to the personalized behaviors of the people to attend the informative needs of reduced targets, at low cost (Cohen et al. 2011).

As noted by Túñez-López et al. (2018), despite the fact that communication flows with audiences are changing and the strengthening of social networks as a new support for transmission and meeting with the public, a new way of symmetrical bidirectional relationship is created, journalists still do not foresee Artificial Intelligence and the automation of contents as an element of transformation or revitalization in the relations of the informants, or the media and with audiences.

> Only one in ten journalists considers that the AI will allow a more personalized relationship with the public, following the same line as the personalization of content that is identified as a differentiating element of the new ways of understanding marketing and organizations' communication managment in this second decade of the 21st century. (Tuñez-López et al. 2018: 756)

Túñez-López, Toural-Bran and Cacheiro-Requeijo synthesized, in 2018, this panorama on a world map of media and companies that use, in a significant way, the automated elaboration of news that is publicly transmitted to which sixteen journalistic media incorporate, thirteen news agencies and 21 companies, mainly from North America, Europe, China and Japan. They also point out that the highest concentrations are in the United States, Germany and the United Kingdom.

In the review of scientific literature, no significant research on the automation of news in Spain has been found, in addition to the contributions of Tuñez-López et al. (2018) and Salazar (2018). Fanta (2017) refers in its study of European agencies that Efe Agency has not yet considered its use, although it points out that some of its delegations do work with small systems of automated data processing.

It was also identified pioneering experiences such as Vocento's when creating service information about beaches or ski resorts, in a project to automate the updating of contents called Medusa; or companies such as Narrativa, that is among the pioneers in the preparation of reports on sports theme in real time for the editorial field and which is already materialized in initiatives such as the newspaper *Sport* to inform the narration of football matches of the Second Division B.

6 The Future: Singling Out the Journalist's Contributions

Artificial Intelligence experts recognize that the challenge is to move from machines programmed to act, to machines with the ability to decide how to act on each occasion: robots with autonomy and ability to think and program their reactions. The exit of the future that is aimed in journalism has not to do with the standardization of texts and texts generated by the machines but to reinforce the contribution of human intervention to the generation of identifiable value in the text through the singularization of proposals or approaches.

That is to say, to reinforce the weight of the cognitive part in the participation of the journalist in the process of construction of the informative agenda, which would suppose a posture that flees from agendas programs, opposing the personal singularization of the informative proposals. Or what is the same, try to dodge the algorithm emphasizing the intellectual component that converts the generation of news in a process away from repetitive mechanical decisions (productive routines) and texts written fleeing clichés, to differentiate them from the repetitive wording in structures and in terminology that currently characterizes the texts produced by the machines.

Acknowledgements This text has been prepared within the framework of *New values, governance, funding and public media services for the Internet society: European and Spanish contrasts* (RTI2018-096065-B-I00) project, of Ministry of Science, Innovation and Universities, co-financed by FEDER; and within the framework of *Digital native media in Spain: storytelling formats and mobile strategy* (RTI2018-093346-B-C33) project, of Ministry of Science, Innovation and Universities, co-financed by the ERDF.

References

Anderson CW (2011) Notes towards an analysis of computational journalism. HIIG discussion paper series 2012(1), pp 1–25

Broussard M (2015) Artificial intelligence for investigative reporting: using an expert system to enhance journalists' ability to discover original public affairs stories. Digit J 3(6):814–831. https://doi.org/10.1080/21670811.2014.985497

Carlson M (2014) The robotic reporter: automated journalism and the redefinition of labor, compositional forms, and journalistic authority. Digit J 3(3):416–431. https://doi.org/10.1080/21670811.2014.976412

Cervera J (2017) El futuro del periodismo es cyborg. Cuadernos de periodistas: revista de la Asociación de la Prensa de Madrid 34:102–109. Retrieved from http://www.cuadernosdeperiodistas.com/futuro-del-periodismo-ciborg

Chu Z, Gianvecchio S, Wang H, Jajodia S (2010) Who is tweeting on Twitter: human, bot, or cyborg?. In: Proceedings of the 26th annual computer security applications conference, Austin, Texas, 6–10 Dec. https://doi.org/10.1145/1920261.1920265

Clerwall C (2014) Enter the robot journalist. J Pract 8(5):519–531. https://doi.org/10.1080/17512786.2014.883116

Cohen S, Hamilton JT, Turner F (2011) Computational journalism. Commun ACM 54(10):66–71. https://doi.org/10.1145/2001269.2001288

Córdoba-Guardado, S (2007) La representación del cuerpo futuro (Ph.D. dissertation). Universidad Complutense de Madrid. Retrieved from https://eprints.ucm.es/7536/1/T29917.pdf

Diakopoulos N (2014) Algorithmic accountability. Digit J 3(3):398–415. https://doi.org/10.1080/21670811.2014.976411

Dickerson JP, Kagan V, Subrahmanian VS (2014) Using sentiment to detect bots on twitter: are humans more opinionated than bots? In: Proceedings of the 2014 IEEE/ACM international conference on advances in social networks analysis and mining, pp 620–627. https://doi.org/10.1109/asonam.2014.6921650

Dörr K (2016) Mapping the field of algorithmic journalism. Digit J 4(6):700–722. https://doi.org/10.1080/21670811.2015.1096748

Edge A (2014) Ophan: key metrics informing editorial at The Guardian. Retrieved from https://www.journalism.co.uk/news/how-ophan-offers -bespoke-data-to-inform-content-at-the-guardian/s2/a563349

Fanta A (2017) Putting Europe's robots on the map: automated journalism in news agencies. In: Reuters Institute for the Study of Journalism. Oxford. Retrieved from https://reutersinstitute.politics.ox.ac.uk/sites/default/files/2017-09/Fanta%2C%20Putting%20Europe%E2%80%99s%20Robots%20on%20the%20Map.pdf

Ferrara E, Varol O, Davis C et al (2016) The rise of social bots. Commun ACM 59(7):96–104. https://doi.org/10.1145/2818717

Flew T, Christina Spurgeon AD, Swift A (2012) The promise of computational journalism. J Pract 6(2):157–171. https://doi.org/10.1080/17512786.2011.616655

Gao T, Hullman JR, Adar E et al (2014) Newsviews: an automated pipeline for creating custom geovisualizations for news. In: Proceedings of the SIGCHI conference on human factors in computing systems, Toronto, Ontario, Canada, April 26–May 1. https://doi.org/10.1145/2556288.2557228

Glahn HR (1970) Computer worded forecasts. Bull Am Meteorol Soc 51(12):1126–1132. Retrieved from https://journals.ametsoc.org/doi/pdf/10.1175/1520-0477%281970%29051%3C1126%3ACPWF%3E2.0.CO%3B2

Graefe A (2016) Guide to automated journalism. Retrieved from https://www.cjr.org/tow_center_reports/guide_to_automated_journalism.php

Graefe A, Haim M, Haarmann B et al (2016) Readers' perception of computer-generated news: credibility, expertise, and readability. Journalism 19(5):595–610. https://doi.org/10.1177/1464884916641269

Gynnild A (2014) Journalism innovation leads to innovation journalism: the impact of computational exploration on changing mindsets. Journalism 15(6):713–730. https://doi.org/10.1177/1464884913486393

Hansen M, Roca-Sales M, Keegan J et al (2017) Artificial Intelligence: Practice and Implications for Journalism. Brown Institute for media innovation and the tow center for digital journalism. Columbia Journalism School. https://doi.org/10.7916/d8x92prd

Hwang T, Pearce I, Nanis M (2012) Socialbots: voices from the fronts. Interactions 19(2):38–45. https://doi.org/10.1145/2090150.2090161

Karlsen J, Stavelin E (2013) Computational journalism in norwegian newsrooms. J Pract 8(1):34–48. https://doi.org/10.1080/17512786.2013.813190

Keeney M (2015) Future cast: will robots replace journalists like toll collectors? Publicity Club of New England. Retrieved from https://www.pubclub.org/837/future-cast-will-robots-replace-journalists-like-toll-collectors

Kim J, Lee K, Kim Y, Kuppuswamy NS, Jo J (2007) Ubiquitous robot: a new paradigm for integrated services. In: 2007 IEEE international conference on robotics and automation, pp 2853–2858

Latar NL (2014) Robot journalists: 'Quakebot' is just the beginning. Knowledge@Wharton, University of Pennsylvania. Retrieved from http://knowledge.wharton.upenn.edu/article/will-robot-journalists-replace-humanl-ones

Latar NL (2015) The robot journalist in the age of social physics: the end of human Journalism? In: Einav G (ed) The new world of transitioned media: the economics of information, communication, and entertainment: the impacts of digital technology in the 21st century. Springer International Publishing, Switzerland, pp 65–80. https://doi.org/10.1007/978-3-319-09009-2_6

Lecompte C (2015) Automation in the newsroom. Nieman Rep 69(3):32–45. Retrieved from http://niemanreports.org/wp-content/uploads/2015/08/NRsummer2015.pdf

Levy S (2012) The rise of the robot reporter. Wired 20(5):132–139

Lindén C (2017) Algorithms for journalism: the future of news work. J Media Innov 4(1):60–76. https://doi.org/10.5617/jmi.v4i1.2420

Lokot T, Diakopoulos N (2016) News bots: automating news and information dissemination on Twitter. Digit J 4(6):682–699. https://doi.org/10.1080/21670811.2015.1081822

Marconi F, Siegman A (2017) The future of augmented journalism: a guide for newsrooms in the age of smart machines. About AP insights. Retrieved from https://insights.ap.org/uploads/images/the-future-of-augmented-journalism_ap-report.pdf

Matsumoto R, Nakayama H, Harada T et al (2007) Journalist robot: robot system making news articles from real world. Paper presented at the 2007 IEEE international conference on robotics and automation, Oct 29–Nov 2. https://doi.org/10.1109/iros.2007.4399598

Meehan JR (1977) TALE-SPIN, An interactive program that writes stories. Int Jt Conf Artif Intell 77(1):91–98. Retrieved from http://citeseerx.ist.psu.edu/viewdoc/summary?doi=10.1.1.74.173

Meyer P (1975) Precision journalism. Commun Inf 1(1):164–165

Mittal S, Kumaraguru P (2014) Broker bots: analyzing automated activity during high impact events on twitter (Ph.D. dissertation). Indraprastha Institute of Information Technology, New Delhi. Retrieved from https://arxiv.org/pdf/1406.4286.pdf

Napoli P (2012) Audience evolution and the future of audience research. Int J Media Manag 14(2):79–97. https://doi.org/10.1080/14241277.2012.675753

Oremus W (2015) No more pencils, no more books. Slate. Retrieved from http://publicservicesalliance.org/wp-content/uploads/2015/10/Adaptive-learning-software-is-replacing-textbooks-and-upending-American-education.-Should-we-welcome-it.pdf

Powers M (2012) "In forms that are familiar and yet-to-be invented": American journalism and the discourse of technologically specific work. J Commun Inq 36(1):24–43. https://doi.org/10.1177/0196859911426009

Renó D, Renó L (2017) Algoritmo y noticia de datos como el futuro del periodismo transmedia imagético. Rev Lat de Comunicación Soc 72:1468–1482. https://doi.org/10.4185/RLCS-2017-1229

Reuters Institute and University of Oxford (2017) Digital news report. Retrieved from http://www.digitalnewsreport.org/survey/2017/

Rivera-Estrada JE, Sánchez-Salazar DV (2016) Inteligencia artificial ¿reemplazando al humano en la psicoterapia? Escritos 24(53):271–291. https://doi.org/10.18566/escr.v24n53.a02

Salazar I (2018) Los robots y la inteligencia artificial. Nuevos retos del periodismo. Doxa Comunicación 27:295–315. https://doi.org/10.31921/doxacom.n27a15

Shearer M, Basile S, Geiger C (2014) Datastringer: easy dataset monitoring for journalists. In: Proceedings of symposium on computation + journalism

Slater MD, Rouner D (2002) Entertainment-education and elaboration likelihood: understanding the processing of narrative persuasion. Commun Theor 12(2):173–191. https://doi.org/10.1111/j.1468-2885.2002.tb00265.x

Starbird K, Leysia P, Hughes A et al (2010) Chatter on the red: what hazards threat reveals about the social life of microblogged information. In: Proceedings of the 2010 ACM conference on computer supported cooperative work, Savannah, Georgia, 6–10 Feb. Retrieved from https://dl.acm.org/citation.cfm?doid=1718918.1718965

Stavelin E (2014) Computational journalism: when journalism meets programming (Ph.D. dissertation). University of Bergen

Steiner T (2014) Telling breaking news stories from Wikipedia with social multimedia: a case study of the 2014 winter olympics. Retrieved from https://arxiv.org/ftp/arxiv/papers/1403/1403.4289.pdf

Strömbäck J (2005) In search of a standard: four models of democracy and their normative implications for journalism. J Stud 6(3):331–345. https://doi.org/10.1080/14616700500131950

Tavares G, Faisal A (2013) Scaling-laws of human broadcast communication enable distinction between human, corporate and robot Twitter users. PLoS ONE 8(7):e65774. https://doi.org/10.1371/journal.pone.0065774

Thurman N, Dörr K, Kunert J (2017) When reporters get hands-on with robo-writing: professionals consider automated journalism's capabilities and consequences. Digit J 5(10):1240–1259. https://doi.org/10.1080/21670811.2017.1289819

Túñez-López JM, Toural-Bran C, Cacheiro-Requeijo S (2018) Uso de bots y algoritmos para automatizar la redacción de noticias: percepción y actitudes de los periodistas en España. El profesional de la información 27(4):750–758. https://doi.org/10.3145/epi.2018.jul.04

Usher N (2017) Venture-backed news startups and the field of journalism. Digit J 5(9):1116–1133. https://doi.org/10.1080/21670811.2016.1272064

Van Dalen A (2012) The algorithms behind the headlines. J Pract 6(5–6):648–658. https://doi.org/10.1080/17512786.2012.667268

Winkler M (2014) The bloomberg way: a guide for reporters and editors. Wiley, New Jersey

Wölker A, Powell TE (2018) Algorithms in the newsroom? News readers perceived credibility and selection of automated journalism. Journalism (Online first, Feb 18). https://doi.org/10.1177/1464884918757072

Young M, Hermida A (2015) From Mr. and Mrs. Outlier to central tendencies. Computational journalism and crime reporting at the Los Angeles Times. Digit J 3(3):381–397. https://doi.org/10.1080/21670811.2014.976409

José Miguel Túñez-López Doctor in Journalism from the Autonomous University of Barcelona and Professor of Organizational Communication and Communication Strategies and Plans, at Universidade de Santiago de Compostela (USC). He is also director of the International Doctoral School of Arts, Humanities, Social and Legal Sciences of the USC and member of the Novos Medios research group (USC). The author has also been Dean of the Faculty of Communication Sciences of the USC (2004–2009) and received the National Queen Journalism Award.

Carlos Toural-Bran Doctor in Communication Sciences from Universidade de Santiago de Compostela. Chief of Staff of the Rector of Universidade de Santiago de Compostela. Associate Professor of Information Architecture, Online Media, DatViz and Organizational Communication. Toural-Bran is also President of AGACOM (Galician Association of Communication Researchers) and secretary of Novos Medios research group. He has also been Vice-dean of the Faculty of Communication Sciences of Universidade de Santiago de Compostela (2014–2019).

Ana Gabriela Frazão-Nogueira Doctor in Audiovisual Communication and Journalism from Universidade de Santiago de Compostela (2016), the author also has the diploma in Advanced Studies in the Communication and Journalism Programme of the Doctorate at the USC (2007–2009). Degree in Communication Sciences from the Fernando Pessoa University, Porto (1995), Ana Nogueira's curriculum also includes, among others, training in Creative Writing, Teaching and Multimedia (2004), Use and Treatment of Voice for New Technologies (2000) and Journalism and Design Techniques of Image (1998).

The Technology-Led Narrative Turn

Jorge Vázquez-Herrero, Xosé López-García and Fernando Irigaray

Abstract The narrative renovation has been a constant throughout history, fed by successive literary and journalistic movements. In the third millennium a new phase starts, with the characteristics of the complexity of the network society and the current technologies as actors in this turn. The definitive rupture of the sequential story has led to experimentation with narrative models based on hypertextuality, multimedia and interactivity. Evolution has been strongly dependant and affected by the past, without major upheavals in the fundamental, although driven by some disruptive dimensions in communication processes. The appearance of journalistic narratives has been guided in these last two decades by the combination of creativity and innovation in an increasingly mobile, convergent and transmedia context. The result is an expansion of models with unequal uses and consumption but enriching the narrative outlook.

Keywords Journalism · Narrative · Multimedia · Convergence

1 Introduction

Social, political and economic contexts have framed the narrative renovation throughout history and have contributed to the construction of the present. The narrative, which simplifying can be understood as the story told by a narrator, is characterized by the combination of these elements in all their complexity and that is what differentiates it from other genres (Valles Calatrava 2008). Techniques linked to temporalization and spatialization, the narrative voice, the narrative speed, the characters or testimonies of actors that intervene in the story are among the most used resources.

J. Vázquez-Herrero (✉) · X. López-García
Universidade de Santiago de Compostela, Santiago de Compostela, Spain
e-mail: jorge.vazquez@usc.es

X. López-García
e-mail: xose.lopez.garcia@usc.es

F. Irigaray
Universidad Nacional de Rosario, Rosario, Argentina
e-mail: fgirigaray@gmail.com

© Springer Nature Switzerland AG 2020
J. Vázquez-Herrero et al. (eds.), *Journalistic Metamorphosis*,
Studies in Big Data 70, https://doi.org/10.1007/978-3-030-36315-4_3

29

They have accompanied the changes in the narrative scenario. Therefore, they also affect the field of journalistic narratives, since the latter are almost always attentive to innovations taking place within the literary scope as a source of inspiration.

The journalistic narrative, as a narrative of constructed reality (Casals 2001), has changed under the influence of the cultural context, with a special impact of film and literature. It has been a valued territory to the members of many journalistic movements, from the so-called New Journalism to the current narrative journalism. It is just this narrative side of journalism the one which has cultivated the writing about certain facts using structures and narrative strategies typical of literature (Herrscher 2012) with better fortune, enriching thus the reality narration. A tradition of promiscuous relationships characterizes the connection between journalism and literature (Chillón 1999), with more or less intensity depending on the historical periods.

Narrators and scholars started this travel when the network society established a new scenario for journalistic stories with the beginning of the third millennium. Now the old models coexist with other renewed ones. Without losing connection with the past, narrative modalities emerge that apply immersive techniques (whether old or new), at the same time the technological dimension opens up little explored or even unknown territories in the past. Looking at and telling reality from narrative journalism (Angulo Egea 2014) has been and is still a challenge. The factual word (Chillón 2014) does not refuse its past relations. It keeps them in the present, but tries new forms and formats that make the dream of a more efficient communication come true.

2 The Digital Incentive

The technological systems that surrounded the birth and evolution of the Internet, produced socially and in a specific cultural environment (Castells 2001), framed the path followed by journalism in the change from the second to the third millennium. The emergence of the network of networks was a real stimulus for journalism, which has compelled professionals to adapt their activity to the new technological scenario. The beginning of journalism on the Internet was the start of a great transformation for the sector, the profession and society. It boosted the characteristics of the network itself and digital technology, which represented a turning point in the design of tools and changes in the processes.

Digitalization, which has come to stay, began its journey with the computerization of production, the introduction of electronic newsrooms, and the digitization of the product (Díaz Noci 2002); even at present, when the digitalization covers everything and places us facing digital citizenship in a data society (Hintz et al. 2019). The first steps of journalistic writing on the Internet include the proliferation of manuals to understand the first consequences of the encounter between journalism and digital technology (Díaz Noci and Salaverría 2003) as well as of analysis and proposals on a new variant of journalistic writing (Salaverría 2005). Since then we have arrived to the application of virtual reality and transmedia strategies in the construction of journalistic discourse, with modalities of reception narratives described as inclusive (Gauthier 2018).

Technology has always been in the construction of modern journalism, from Gutenberg's typographical invention to the subsequent steps to the industrial age, when it has lived under the influence of industrial society (Gómez Mompart and Marín Otto 1999). Since the appearance of computerization, which changed the newsrooms, technology gained presence and precision techniques, with current tools and the scientific method (Meyer 1993). Later, from the first steps of digital journalism, the impact of technology influenced how journalists do their work, the content of the news, the structure and organization of the newsrooms, and the relationships between journalism organizations, professionals and its users (Pavlik 2010). The consequences have affected the set of processes surrounding journalistic work, both the creation of content, with the opening of new ways in the narratives, as well as its distribution.

Online journalism has increased the complexity of the different dimensions of this social communication technique, especially through the construction of collaborative narratives with different management models of citizen participation. Since the emergence of multimedia narratives (Deuze 2004) we have entered new territories such as transmedia narratives (Scolari 2009) or immersive journalism (De la Peña et al. 2010). The processes of journalistic innovation (Paulussen 2016) look for answers to the challenges, with special attention to the narrative field. Formats and techniques are multiplied to seek a kind of communication that takes better advantage of the possibilities of the current ecosystem and journalistic pieces that reach a more efficient communication. The result is the existence of a great diversity of proposals, models, currents and formats that enrich the journalistic narratives in the prelude of the third decade of the 21st century.

3   Multimedia, Interactive and Immersive Scenario

The arrival of the computer and the Internet to the newsrooms demanded changes in the nineties. The first online media were established in a digital scenario where the rules of the analogue world no longer worked (Pavlik 2001); a scenario marked by the convergence in production and consumption (Deuze 2007), where the rigid classifications of the media (press, television, radio) had become obsolete. In a first phase, the hypertextual condition was implemented, the most unique feature of the World Wide Web. In practice, it meant creating connected documents, which would be enriched by multimedia construction with the combination of resources such as photography, video or infographics together with text.

The software becomes fundamental during this stage in the creative and informative processes (Manovich 2014). Among the most relevant tools were the Macromedia and Adobe products, making the edition and multimedia production more accessible with their specific applications. In the 21st century, with the appearance of content management systems such as Drupal, Wordpress or Joomla, the publication of digital media became more affordable. These platforms are adaptable to the needs of each project and they greatly facilitated the design of the websites and their management. However, the standards are the ones that really facilitate and consolidate new functionalities in digital products. The most accessible channel for the public is the Web, thanks to the existence of standardized conditions that allow a homogeneous visualization of the contents and without requiring devices of the latest technology, just one connected to the network. The implementation of HTML5, CSS3 and Javascript has facilitated the creation of multimedia projects for the Internet with improved integration of video, animations and interactive and personalized complements.

However, specialized tools emerge as a response to the complex requirements for multimedia and interactive production. These are promoted by producers, media and consultants, sometimes with funding from innovation plans such as Google's Digital News Innovation Fund. Klynt is the interactive documentary creation software of the French studio Honkytonk, Conducttr is a mixed reality platform created by Robert Pratten, and Shorthand is the tool used by multiple media and agencies for multimedia publications. These applications, among many others, are being used to provide interactivity, visualize data, generate new formats or combine resources in a fluid and attractive way.

The so-called multimedia journalism is a contemporary way of telling stories (George-Palilonis 2012), "a critical middle ground between the impenetrable overloads and binary simplifications of digital communication" (Ball 2016: 432). In this sense, the paradigmatic report of *The New York Times* published in 2012, *Snow Fall: The Avalanche at Tunnel Creek*, stood out for the integration of multimedia resources and the fluidity of its scroll navigation. This special coverage was developed in HTML with jQuery plugins that provided dynamism. It has consolidated a tendency for the reports with visual prominence and *scrollytelling*. But in addition to being an alternative to the accelerated consumption of information, multimedia journalism opens up new options to reach readers and create conversation (Lassila-Merisalo 2014).

Among multimedia formats there is a commitment to visual content, photography and video, which occupy most of the screen. The media faces the challenge of representing data in accessible ways, the revaluation of audio and the integration of resources in the same design. However, this did not mean great progress in the last decade. Again, technological advances determine innovative formats. As we will see next, these point to interactivity and immersion as already developed factors while automation is still developing.

3.1   The Interactive Promise

From the very definition of online journalism, interactivity (Deuze 2003; Rost 2006) has been the differential characteristic and the eternal promise. On an instrumental and navigational level, interactivity has had a wide implementation in the media;

however, it is more incomplete in relation to participation. The ability to generate bidirectional communication (Gershon 2016) has barely reached an exploratory level, mainly through comments on news (Ziegele et al. 2014) and online videos (Ksiazek et al. 2016). However, the potential of interactivity hasn't exploded in the media yet, also limited by the risks of a communication shared by the audience; an uncomfortable myth (Domingo 2008) for the newsrooms with consolidated production routines.

The development of devices has favoured the evolution of interactivity, especially through mobile devices due to their constant connection to the network. Platforms condition the content, although the media moves towards an indifferent treatment of the journalistic content according to the platform (Westlund 2013). Other factors that reinforce it come into play. Social networks and their popularization are the main platforms for the articulation of interactivity in the media. In interactive works, its design becomes the motor of interaction, as a trigger and facilitator of the experience. In addition, changes in consumer behaviour and the existence of an active user profile support the emergence of interactive projects.

Different formats have risen after the bet of the media for interactivity, with an experimental nature and a significant production cost as well. Interactive documentary is a good example. It is a format that gives greater control to the user, who can optionally diverge from the central story to focus their attention on specific aspects, consume personalized content or contribute with their own content to the expansion of the project (Vázquez-Herrero and Gifreu-Castells 2019). Gamification is also a strategy present in the media, which implies the adoption of game elements in contexts that are not ludic (Deterding et al. 2011), giving rise to specific formats such as newsgames, serious games or docugames. In the gamified projects developed by the media, the user assumes a role and gets journalistic information in interviews and data visualizations.

At a higher level in terms of interactivity, participatory and collaborative projects are found, although with a very scarce presence in the field of news media. These consider the participation of an active audience, a willingness to contribute that is favoured by the technological developments that bring people closer through digital media and really enable interaction.

Interactivity as a possible condition –and also in exploration– allows a wide range of opportunities, but struggles with the structures and routines of the media organizations, the high cost of creating the most sophisticated systems and a doubtful economic return. There are fears, uncertainties and difficulties, but it is still a premature field where "journalists have only scratched the surface of digital affordances for storytelling" (Murray 2017: 77).

3.2 Immersion in Reality

Since 2010, the concept of immersive journalism has been developed (De la Peña et al. 2010; Domínguez 2013) supported mainly in virtual reality and video games. It is one of the most recent technological change that directly affects journalistic narratives.

habits are indispensable conditions for the production of adapted narratives. Something which allows us to understand why convergence is not a purely technological process. Rather, as a cultural change, it encourages consumers to seek information and establish connections between dispersed media content. Convergence occurs in the consumers' brains and in their social interactions. Therefore, it is a process that develops in the cultural layer and not only in the technological one, where the circulation expands and transcends the media, jumping between platforms where the universe of the narrative framework unfolds.

In a hyperconnected, dispersed and transmedia context, narrative strategies can use multiple platforms to expand their content, opening to the participation of users who seek to be protagonists. Jenkins (2008) says that, in the culture of convergence, consumers are increasingly powerful because they are learning to participate and interact in new environments. The narrative strategies in the paradigm of convergence face a double challenge: producing autonomous texts with feedback and multiple formats, for analogue and digital platforms, giving rise to the participation of users in the construction of knowledge and the narrative universe.

Audiences today are participatory, and it is one of the most tangible effects of the irruption of a new technological scenario that transformed the conditions of production, circulation and consumption of information. In this context, the use and appropriation of the new digital connection devices had an impact on the traditional models of the communication field, which begins to discuss new categories of analysis.

For Jenkins (2008), these changes in the media paradigm are crossed by three central concepts: media convergence, participatory culture and collective intelligence. In his definition, Jenkins does not conceive convergence as a merely technological process, its meaning is broader as it operates in a cultural dimension:

> With convergence I refer to the flow of content across multiple media platforms, the cooperation between multiple media industries and the migratory behaviour of media audiences, willing to go almost anywhere in search of the desired type of entertainment experiences. (Jenkins 2008: 14)

New audiences are fragmented, collaborative and interactive from the use of a multiplicity of media and languages for the production, distribution and consumption of content on the network and they are no longer subject to messages built by traditional media. Media convergence alters the relationship between existing technologies, markets, genres and the public. The power of the producer and media consumers interacts in unpredictable ways. This implies a change of paradigm where the audiences multiply, with differentiated needs and pretensions that the mode of the broadcast does not reach to satisfy, fundamentally changing thus the distribution criterion, which leads to a continuous update model.

A transmedia project can take many forms and aspects, but always taking into account "optimizing the advantages that transmedia gives us, in the possibility of delivering the right content for the right device at the appropriate time" (Pratten 2011: 6). The participation of the experience is as important as the way you tell the story. You have to make an effort to remember that the emotional engagement is achieved by the combination between a good story and the design of the experience that will be submitted to the audience. For this, it is essential to take into account four factors that intervene in the design: the platforms, the defined spaces, the synchronization of the contents and the participation actions that include calls to action.

With the multiplication of screens, users' capabilities to fragment their attention in several platforms simultaneously were increased in equal measure. A fundamental characteristic of this new way of storytelling is that different stories change the equation in the development of the storyword, by using different languages to build autonomous pieces in the narrative framework. These are "intertwined, without overlaps and keeping their independence from the others" (Maguregui 2010: 107).

Offering multiple points of view of the narrative world enriches the user's experience and serves not only to increase the number of works related to the same story, but also to provide different entry points. For this process to really take place, the user's experience must have the same importance as narrative development at design stage. Unfortunately, it is something generally left to chance by content producers, who begin to think about these developments but actually think about the story only.

5 Conclusion

The evolution of technology has largely determined the forms to communicate. Historically, the incorporation of new systems such as printing, radio or television have shaped the communicative processes, creating a balance between the adaptation of the known and the experimentation of the possibilities that are opened. In recent decades, narratives have mutated constantly influenced by technology. The way of telling stories is also affected by a mobile and convergent scenario of multiple platforms.

Journalism is moving through a period of rapid and far-reaching changes (Franklin 2016). Although a large part of them are due to the technological effect on production, distribution, consumption or business models, convergence at the same time shows cultural and social changes. The user's behaviours turn towards a more participatory and interactive communication in a context of interstitial leisure and, above all, mobile. For this reason, ubiquity, transmediality and micro-contents define emerging narratives, with a strong link with social media and their associated habits.

Technologies have not only offered opportunities for journalism, but also challenges, and have required strategic decisions for innovation and experimentation. 5G connectivity, immersive audio, augmented reality and narrative automation are currently areas of exploration to continue telling stories, once again, with technology as inspiration to rethink old definitions.

Acknowledgements This article has been developed within the research project *Digital native media in Spain: storytelling formats and mobile strategy* (RTI2018-093346-B-C33) funded by the Ministry of Science, Innovation and Universities (Government of Spain), Agencia Estatal de Investigación, and co-financed by the European Regional Development Fund (ERDF), and *Narrativas inmersivas: Realidad virtual y realidad aumentada en relatos de no ficción* (POL274, Facultad de Ciencia Política y RR. II., Universidad Nacional de Rosario), as well as it is part of the activities promoted by Novos Medios research group (ED431B 2017/48), supported by Xunta de Galicia. The author Jorge Vázquez-Herrero is a beneficiary of the Faculty Training Program funded by the Ministry of Science, Universities and Innovation (FPU15/00334).

References

Amar G (2011) Homo mobilis: la nueva era de la movilidad. La Crujía, Buenos Aires

Angulo Egea M (2014) Prefacio. Mirar y contar la realidad desde el periodismo narrativo. In: Angulo M (ed) Crónica y mirada. Aproximaciones al periodismo narrativo. Libros del KO, Madrid

Ball B (2016) Multimedia, slow journalism as process, and the possibility of proper time. Digit J 4(4):432–444. https://doi.org/10.1080/21670811.2015.1114895

Casals MJ (2001) La narrativa periodística o la retórica de la realidad construida. Estudios sobre el Mensaje Periodístico 7:195–219

Castells M (2001) La Galaxia Internet. Reflexiones sobre internet, empresa y sociedad. Plaza & Janes Editores, Barcelona

Chillón A (1999) Periodismo y literatura. Universitat Autònoma de Barcelona, Barcelona, Una tradición de relaciones promiscuas

Chillón A (2014) La palabra fáctica. Literatura, periodismo y comunicación. Universitat Autònoma de Barcelona, Barcelona; Universitat Jaume I, Castelló de la Plana; Universitat Pompeu Fabra, Barcelona; Universitat de València, Valencia

De la Peña N, Weil P, Llobera J, Giannopoulos E, Pomés A, Spanlang B, Friedman D, Sánchez-Vives MV, Slater M (2010) Immersive journalism: immersive virtual reality for the first-person experience of news. Presence 19(4):291–301

Deterding S, Dixon D, Khaled R, Nacke L (2011) From game design elements to gamefulness: defining "gamification". In: Proceedings of the 15th international academic MindTrek conference: envisioning future media environments. ACM, Tampere, pp 9–15. https://doi.org/10.1145/2181037.2181040

Deuze M (2003) The web and its journalisms: considering the consequences of different types of newsmedia online. New Media Soc 5(2):203–230. https://doi.org/10.1177/1461444803005002004

Deuze M (2004) What is multimedia journalism? J Stud 5(2):139–152

Deuze M (2007) Convergence culture in the creative industries. Int J Cult Stud 10(2):243–263. https://doi.org/10.1177/1367877907076793

Díaz Noci J (2002) La escritura digital. Hipertexto y construcción del discurso informativo en el periodismo electrónico. Universidad del País Vasco, Bilbao

Díaz Noci J, Salaverría R (2003) Manual de redacción ciberperiodística. Ariel, Barcelona

Domingo D (2008) Interactivity in the daily routines of online newsrooms: dealing with an uncomfortable myth. J Comput Med Commun 13(3):680–704. https://doi.org/10.1111/j.1083-6101.2008.00415.x

Domínguez E (2013) Periodismo inmersivo. La influencia de la realidad virtual y del videojuego en los contenidos informativos. Editorial UOC, Barcelona

Engberg M, Bolter JD (2014) Cultural expression in augmented and mixed reality. Convergence 20(1):3–9. https://doi.org/10.1177/1354856513516250

Franklin B (2016) The future of journalism. J Stud 17(7):798–800

Future Today Institute (2017) 2018 tech trends for journalism and media. Retrieved from https://futuretodayinstitute.com/2018-tech-trends-for-journalism-and-media. Accessed 1 Apr 2019

Gauthier P (2018) Inmersion, social media and transmedia storytelling: the "inclusive" mode of reception. Comun Med 37:11–23. https://doi.org/10.5354/0719-1529.2018.46952

García Marín D (2017) La nueva comunicación sonora. Del podcast al transcasting. In: Aparici R, García Marín D (eds) ¡Sonríe, te están puntuando! Narrativa digital interactiva en la era de Black Mirror. Gedisa, Barcelona, pp 145–163

George-Palilonis J (2012) The multimedia journalist. Storytelling for today's media landscape. Oxford University Press, Oxford

Gershon RA (2016) Digital media and innovation: management and design strategies in communication. Sage, Los Angeles

Gómez Mompart JL, Marín Otto E (1999) La irrupción de la información televisiva y la influencia del periodismo singular. In: Gómez Mompart JL, Marín Otto E (eds) Historia del periodismo universal. Editorial Síntesis, Madrid

Google News Lab (2017) Storyliving: an ethnographic study of how audiences experience VR and what that means for journalists. Retrieved from https://newslab.withgoogle.com/assets/docs/storyliving-a-study-of-vr-in-journalism.pdf. Accesses 1 Apr 2019

Herrscher R (2012) Periodismo narrativo. Cómo contar la realidad con armas de la literatura. Publicacions y Edicions de la Universitat de Barcelona, Barcelona

Hintz A, Dencik L, Wahl-Jorgensen K (2019) Digital citizenship in a datafied society. Polity Press, Cambridge

Igarza R (2009) Burbujas de ocio. Nuevas formas de consumo cultural. La Crujía, Buenos Aires

Igarza R, Vacas F, Vibes, F (2008) La cuarta pantalla. Marketing, publicidad y contenidos en la telefonía móvil. Lectorum-Ugerman, Buenos Aires

Irigaray F (2016) DocuMedia: documentales multimedia interactivos en la periferia. El caso Calles Perdidas. In: Luchessi L, Videla L (eds) Desafíos del periodismo en la sociedad del conocimiento. Viedma, Editorial UNRN, pp 65–78

Jenkins H (2008) Convergence culture. La cultura de la convergencia de los medios de comunicación. Paidós, Barcelona

Ksiazek TB, Peer L, Lessard K (2016) User engagement with online news: conceptualizing interactivity and exploring the relationship between online news videos and user comments. New Media Soc 18(3):502–520. https://doi.org/10.1177/1461444814545073

Lassila-Merisalo M (2014) Story first. Publishing narrative long-form journalism in digital environments. J Mag New Media Res 15(2):1–15

Maguregui C (2010) Cruce de plataformas, arquitectura de la anticipación y régimen de indentificaciones en Lost. In: Maguregui C, Piscitelli A, Scolari C (eds) Lostología: estrategias para entrar y salir de la isla. Cinema, Buenos Aires, pp 101–132

Manovich L (2014) Software is the message. J Vis Cult 13(1):79–81. https://doi.org/10.1177/1470412913509459

Meyer P (1993) Precision journalism. A reporters guide do social science methods. Indiana University Press, Bloomington

Murray JH (2017) Hamlet on the holodeck: the future of narrative in cyberspace. The MIT Press, London

Paulussen S (2016) Innovation in the Newsroom. In: Witschge T, Anderson CW, Domingo D, Hermida A (eds) The Sage handbook of digital journalism. Sage, London

Pavlik J (2001) Journalism and new media. Columbia University Press, New York

Pavlik J (2010) The impact of technology on journalism. J Stud 1(2):229–237. https://doi.org/10.1080/14616700050028226

Pratten R (2011) Getting started with transmedia storytelling. A practical guide for beginners. Createspace, London

Rost A (2006) La interactividad en el periódico digital. Ph.D. dissertation. Universidad Autónoma de Barcelona, Barcelona. Retrieved from https://ddd.uab.cat/pub/tesis/2006/tdx-1123106-104448/ar1de1.pdf. Accessed 1 Apr 2019

Salaverría R (2005) Redacción periodística en internet. EUNSA, Pamplona

Sánchez Laws AL (2017) Can Immersive journalism enhance empathy? Digit J [Online first, 20 Oct 2017]. https://doi.org/10.1080/21670811.2017.1389286

Scolari CA (2009) Transmedia storytelling: implicit consumers, narrative worlds, and branding in contemporary media production. Int J Commun 3:586–606

Vacas F (2010) La comunicación vertical. La Crujía, Buenos Aires

Valles Calatrava JR (2008) Teoría de la narrativa. Una perspectiva sistémica. Iberoamericana, Madrid

Vázquez-Herrero J, Gifreu-Castells A (2019) Interactive and transmedia documentary: production, interface, content and representation. In: Túñez-López M, Martínez-Fernández VA, López-García X, Rúas-Araújo X, Campos-Freire F (eds) Communication: innovation & quality. Springer, Cham, pp 113–127. https://doi.org/10.1007/978-3-319-91860-0_8

Watson Z (2017) VR for news: the new reality? Reuters Institute for the Study of Journalism, University of Oxford. Retrieved from https://reutersinstitute.politics.ox.ac.uk/our-research/vr-news-new-reality. Accessed 1 Apr 2019

Westlund O (2013) Mobile news. digital. Journalism 1(1):6–26. https://doi.org/10.1080/21670811.2012.740273

Ziegele M, Breiner T, Quiring O (2014) What creates interactivity in online news discussions? an exploratory analysis of discussion factors in user comments on news items. J Commun 64:1111–1138. https://doi.org/10.1111/jcom.12123

Jorge Vázquez-Herrero Universidade de Santiago de Compostela, Spain. Ph.D. in Communication, Universidade de Santiago de Compostela (USC). He is a member of Novos Medios research group (USC) and the Latin-American Chair of Transmedia Narratives (ICLA–UNR, Argentina). He was visiting scholar at Universidad Nacional de Rosario, Universidade do Minho, University of Leeds and Tampere University. His research focuses on digital interactive narratives of non-fiction—mainly interactive documentary, micro-formats and transmedia, immersive and interactive narratives in online media.

Xosé López-García Universidade de Santiago de Compostela, Spain. Professor of Journalism at Universidade de Santiago de Compostela (USC), Ph.D. in History and Journalism (USC). He coordinates the Novos Medios research group. Among his research lines there is the study of digital and printed media, analysis of the impact of technology in mediated communication, analysis of the performance of cultural industries, and the combined strategy of printed and online products in the society of knowledge.

Fernando Irigaray Universidad Nacional de Rosario, Argentina. Director of the MA in Interactive Digital Communication, and the Multimedial Communication Department at Universidad Nacional de Rosario (Argentina). He is also Executive Director of the Latin-American Chair of Transmedia Narratives at the Institute of Latin-American Cooperation (ICLA–UNR). He is a director and producer of TV programmes and documentaries, and interactive and transmedia works. He won in 2013 the International Award for Journalism Rey de España in the Digital Journalism category with his team.

Journalism Innovation and Its Influences in the Future of News: A European Perspective Around Google DNI Fund Initiatives

Ana Cecília B. Nunes and João Canavilhas

Abstract Innovation for journalism has become an increasing matter of survival in the new digital media landscape. However, we still face a lack of a shared understanding of media innovation within academia combined with an inability to clearly define what is innovation related to the news industry. This paper contributes to the necessity of better defining journalism innovation by analyzing seventeen highlighted projects on *Google DNI Fund 2018 Report* initiative. We propose a framework of five variables: implementation stage; innovation mindset; innovation focus/innovation target; proven journalist outcomes; and potential replicability degree. The conclusions show that news media innovators are searching for ecosystem solutions rather than worry about solutions for themselves. News business models and news narratives or formats are strongly connected, which means that media might finally understand that they should work for real users and not for imagined ones.

Keywords Journalism innovation · Media innovation · Digital journalism · Google

1 Introduction

The development and spread of technology, as well as its consequent media market changes, has turned innovation into a central theme for news industry survival. Innovation became even more crucial when new incumbents started to steal significant profits from legacy media groups, a pattern described by Bower and Christensen (1995) as the disruption theory. Christensen et al. (2012) comment that what is happening today in the news industry is not much different than what has been seen in other markets, with new actors entering and taking advantage of their digital born nature, without a legacy to worry about. The fact is it disrupted the media market, running all news industry into a no-way-back innovation road. As Küng states: "the

A. C. B. Nunes
Pontifical Catholic University of Rio Grande do Sul, Porto Alegre, Brazil
e-mail: ana.nunes@pucrs.br

J. Canavilhas (✉)
University of Beira Interior, Covilhã, Portugal
e-mail: jc@ubi.pt

© Springer Nature Switzerland AG 2020
J. Vázquez-Herrero et al. (eds.), *Journalistic Metamorphosis*,
Studies in Big Data 70, https://doi.org/10.1007/978-3-030-36315-4_4

requirement for innovation in the media industry has become both more urgent and more challenging as the pace and scope of technological advance have increased" (2013: 9).

Despite media innovation becoming a decisive topic, its conceptualisation is still very diffuse (Dogruel 2013). Academic research on this area experimented a significant growth over the past 10 years (García-Avilés et al. 2018), but we need still "innovation-oriented journalism research that provides clear, foundational definitions of 'innovation' in reference to journalism" (Posetti 2018: 12).

Considering these premises, this paper proposes a discussion around journalism innovation through the analysis of seventeen *Google Digital News Innovation Fund* initiatives highlighted in its three-year report. The Fund is a European project with a €150 million commitment to support and kick-start innovations targeting to help journalism growth in the digital age. The projects are divided into four categories: (1) battling misinformation, (2) telling local stories, (3) boosting digital revenues and (4) exploring new technologies (Google 2018).

Based on those projects, we argue on the innovation types (Lindmark et al. 2013; Storsul and Krumsvik 2013), innovation aims, replicability degree and its outcomes (García-Avilés et al. 2018), proposing a framework to better understand journalism innovation.

The analysis sets ground to discuss what innovation in the news industry could really mean, based on the reflection of Lindmark et al. on the difficulty in defining media innovation: "Where to draw the line between media innovation and routine media production is not obvious" (2013: 130). We seek to identify the main aspects from the innovations focused on a medium to long-term impact, such as the majority of those analyzed here.

The need to understand journalism innovation particularities and, indeed, to propose a clear concept of what it is and how it has been addressed by different social actors to influence the future of the news industry is not exclusively an academic and theoretical demand. It is also a need to the professional field to develop more innovative initiatives in the market industry.

2   Innovation in Media: The Need of a Particular Perspective in a Digital Environment

Since the digital expansion, media landscape has faced a growing tension between the so-called legacy media and the digital-born operations, usually being seen as the most innovative organizations. This fragmented scenario with an increased competition market is not the only factor that brought light to innovation within media, but it is a major topic. If digitalization made possible a spreadable media context (Jenkins et al. 2018), it also brought sustainability issues, as the need for new business models and, among other factors, an urge to understand and better develop innovation for the media and journalism sector.

Indeed, innovation for the media and journalism has not been a popular or even a decisive research topic for media studies before the digital disruption. Lavine and Wackman (1988) are among the first researchers studying challenges regarding media management before the 2000s. The authors highlighted that the media sector had particularities, especially because of the perishable nature of the media product, the highly creative members of the profession, the more flexible and horizontal media organizational structure, in addition to the social role of media and the blurred boundaries that separate traditional media. Ferguson (1991) added that the ubiquitous nature of media, its high visibility, along with the lack of an expertise in the audience's vision, the fact that media industry deals with creativity and moreover, how the managers' work has influence in the gatekeeping process, working as a discourse filter to society, are particular and important characteristics to consider when focusing on media innovation.

Recently, more researchers contributed with new approaches into this subject. Storsul and Krumsvik (2013) revised the types of media innovation by adapting the four axis from the *Oslo Manual* (OECD 2005): product (creation or improvements in the new product), process (new methods around the news work, both internal or external from the newsroom), position (how the product is positioned or framed) and paradigmatic (includes changes in an organization's mindset, values and business models). Storsul and Krumsvik (2013) also added a fifth category: social innovation. According to them, media can use existing products or services in an innovative manner with the major target on pursuing social goals. A fact that could be especially connected with the values of journalism, an activity related with the role of journalism in the democratic societies (Kovach and Rosenstiel 2014).

Those types of innovation can also be related to a more specific media and journalism innovation target, which is where the contribution of Lindmark et al. (2013) takes a slightly different approach than the previous authors. According to them, media innovations should be classified in the realm of content (innovations in the message itself or in a new narrative form), consumption (new ways of consuming content), production and distribution (changes in how to produce, reproduce, distribute or display content) and business model (new business models including new forms of industry organization). The two perspectives are not contradictory; instead, they hold a close relation (Fig. 1).

Despite being possible to relate all Lindmark's et al. (2013) classifications with Storsul and Krumsvik's (2013) one, the opposite does not hold true. In fact, Lindmark et al. (2013) approach appears to be much more related to the innovation aims than García-Avilés et al. (2018). This research work is based on the views of journalists who are leading innovation in Spain and produces a model of innovation diffusion in media outlets. If media is particular regarding innovation types, the work of García-Avilés et al. (2018) shows it also holds true considering innovation diffusion within newsrooms (Fig. 2).

It is perceived a close connection between the innovation types (Lindmark et al. 2013) and the aim of the innovation, emerging from the mentioned model. Although with similarities with the previous concepts discussed, four axis of this proposed

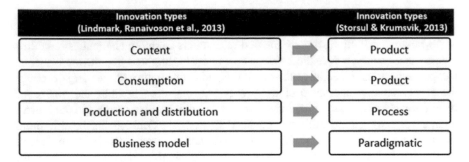

Fig. 1 Lindmark et al. (2013) and Storsul and Krumsvik (2013). Own elaboration

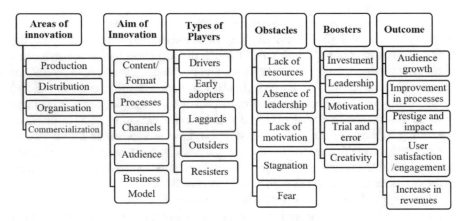

Fig. 2 A model of the diffusion of innovations in media outlets. García-Avilés et al. (2018)

model shine new lights on journalism innovation. They are: types of players, obstacles, boosters and outcomes. As Lavine and Wackman (1988) have stated before, media outlets are unique and that is why the particularities of the journalism innovation constitutes a theme deserving further thoughts.

This research also endorses the relation between competitiveness and innovation within the news industry. Based on the interviews conducted with newsroom leaders in Spain, "innovation involves the capacity to respond to opportunities and threats within the market, thereby 'managing to beat competitors, identify opportunities and take risks', as one editor put it" (García-Avilés et al. 2018: 6).

Finally, among those axis proposed, outcomes captured our attention. Proposed categories in this axis still somehow subjective, as, for instance, prestige and impact. Those two topics could be difficult to measure, especially in a systematic way and in a newsroom environment. In addition, possible outcomes are strongly related with business goals, on the realm of the organization itself, but not considering the news or media future as a whole. This can mean that the possibility of this innovation to spread to other cities, regions or countries, influencing the entire journalism market is not something strongly pursued within newsrooms innovation leaders, or perhaps, something that journalists still do not dare to dream of in the newsroom environment.

To target this issue we need to add a new dimension to the innovations in media and journalism, that is: replicability degree, or whether this innovation could be used in other settings and how difficult it would be to implement or adopt it beyond the sphere of its original environment. It also relates to how the resulted innovation could inspire others or be replicable in other media and/or journalism companies and/or other contexts. In our view, replicable innovations are those that can be appropriated (used with or without modifications) in similar contexts for equivalent or close results. Indeed, the important role of the press in the proper functioning of democracy depends on its capacity for innovation as a whole and not on isolated success case studies, such the importance of innovation replicability to the media ecosystem.

The outcome of media and journalism innovation in a future media landscape can be of great impact in solving digital propagation challenges and for the continuity of a sustainable news environment. Therefore, one question remains: in what degree have innovations around media and journalism been addressing the future and perpetuity of the news industry? This is a topic that certainly deserves further thoughts, and something that we would like to start to address in the following sections of this paper.

3 Innovation in Journalism and Its Impact in the News Industry

One of the main findings of the 2018 *Reuters Innovation in Journalism* report (Posetti 2018) is that the search for innovation within the news industry is so much attached to embrace new technology that it forgets about its purpose or long-term strategy. Without a navigation plan, the sailor could never reach its destination. It does not mean that journalism should not embrace technology or, for instance, that Snapchat Stories should not be tested by newsrooms. It only means that short-term solutions could bring us small and immediate results, usually in the realm of a unique and particular company, but it might be the medium and long-term innovation approaches that could be a game changer for the industry as a whole. As Posetti states, "there is evidence of an increasingly urgent requirement for the cultivation of sustainable innovation frameworks and clear, longer-term strategies within news organisations" (Posetti 2018: 7). It brings us again to the discussion around outcomes, done previously. It is a pure example that the industry might still be thinking as an isolated news business unit instead as of a tile in a media ecosystem. Clearly, it will be just the interconnected innovation thinking that will help us to sail in this turbulent digital business industry waves. "In the absence of purposeful strategy and reflective practice, ad hoc, frantic, and often short-term experimentation is unlikely to lead to sustainable innovation or real progress" (Posetti 2018: 8).

In this scenario, replicability should be an important topic for media and journalism innovations, as it is tied to a long-term approach and a vision of field or group, instead of prioritizing a particular and immediate need inside a single business strategy. We argue they are different, tough, from reproducible journalism or innovations. Reproducible journalism is tied to data and its availability as discussed in a dedicated panel at *Stanford's Computation + Journalism Symposium* in 2016 (Christensen 2016): "This is the question of how data journalism can be reproducible–how should journalists deal with the question of anonymous sources, or leaked data, etc.?". Different from reproducible innovations, replicable ones are flexible, and can impact more diverse contexts with possibilities to go beyond the mere innovation itself, inspiring new innovative propositions, leaving room for adaptations.

In data journalism, a program that allows the recombination of several datasets is a project with a high degree of reproducibility. An example of a project that fits this model is Stacked Up, created by Meredith Broussard, Pam Selle and Jeff Frankl, in 2013[1]. This python-based automation on annual school ranks displays an analysis of the ranking of the best schools in the State of Philadelphia, in the United States, with the number of books in libraries (especially those used in the ranking of educational institutions in the region). The result is the indication of the correlation between those two data points, inferring the need for investment to put schools with fewer books per student better placed on the list. The result of innovation is the automation of the relationship between these disperse elements through the python programming language. It is an interesting initiative, but much different from the *Blendle* business model (The Netherlands), an example of replicable innovation. The latter initiative may inspire others in different contexts with problems of monetization of digital journalism with similarities to the Dutch context.

The difference between using the term reproducible and replicable, we argue, is in fact a matter of expectation and emphasis. Hence, reproducible innovations should tend to have similar results in similar contexts, as the example above. Replicable ones, otherwise, usually emphasizes an output concept with a longer-term view that could be tested in different context, but more important, that could inspire others to also think the same way, not exactly as it was, but as adaptations and replications of its experience.

The fact is that the news business and work environment itself did not contribute historically to long-term thinking. Journalism and journalists have always been dealing with the next deadline or big news-reporting feature, focusing on a short-term plan rather than on a medium or long-term approach. This thinking, however, brought us to a scenario where journalism is hardly prepared to strategic long-term mindset.

Regardless of being a short, medium or long-term solution, many of the innovation initiatives within journalism are nowadays born from interdisciplinary teams, often out of the newsroom daily routines, as many examples analyzed further. "It is acknowledged that 'longer-term strategies' in an industry prone to 'pivoting' in

---

[1]The project resulted in a feature in The Atlantic, written by Broussard: https://www.theatlantic.com/education/archive/2014/07/why-poor-schools-cant-win-at-standardized-testing/374287/

response to relentless change is more likely to be understood as 6–24 months, rather than 5–10 years" (Posetti 2018: 8).

4   Research Methods and Google DNI Fund

Considering the previous discussions, this paper seeks to identify particularities of journalism innovation in the European context, through the study case of *Google DNI Fund* initiatives highlighted in its three-year report.

The referred investment competition is a program with a "€150 million commitment to support and kick-start innovation within the European news ecosystem" (Google 2018). It was chosen as a way to comprise a quite dispersed topic as innovation, addressing its characteristics and proposing a preliminary framework for innovation within the news industry.

The projects presented in the report were highlighted between 461 funded initiatives within its history and, hence, it is understood that those were the most successful ones according to *Google News*. All the seventeen projects were analyzed, based on axis emerged from previous literature review and from the investigation of the initiatives itself (Fig. 3).

The first axis—*Implementation Stage*—emerged from the different innovation phases comprised by the projects: in development, prototype, testing phase or adoption.

The second—*Innovation Mindset*—is the result of the previous discussion regarding innovation impact and the same holds true to potential replicability degree. The level may vary between ecosystem impact (implemented actions towards an industry-wide effect), ecosystem view (innovation cases without any implemented ecosystem action, although ambitioning an industry-wide influence) and unitary impact or view (cases of internal innovation without direct or planned impact outside the companies' sphere or presenting any prospect or evidence of ecosystem view).

The third axis is a new typification of innovation—*Innovation focus/Innovation target*—based on the combination of those three researches (Storsul and Krumsvik 2013; Lindmark et al. 2013; García-Avilés et al. 2018) and are organized in six areas: news content—narratives, formats or niche news content; journalism production or processes; channels of news distribution; audiences participation and community engagement; new business models or impact in journalism sustainability; social—related to social benefits and with journalism democratic goal. A remark needs to be done: the innovation focus/innovation target here refers to the innovation impact direct target, and not the innovation process or innovation development. It means that the community could be very engaged in the product development, but once the innovation was out, in the adoption phase, if it is not directly impacting audience engagement, it was not counted as innovation focus/innovation target for this initiative.

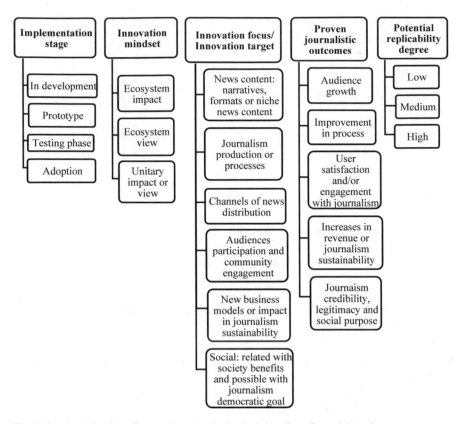

Fig. 3 Proposed framework to analyze innovation in journalism. Own elaboration

The fourth axis—*Proven journalist outcomes*—was inspired in the work of García-Avilés et al. (2018), which interviewed journalists in Spain leading innovation processes in newsrooms. To avoid subjectivity, only the journalistic outcomes that had some data indication of success were considered, that is: those presented by numbers and/or data in each initiative description.[2] For example, a rise in journalism subscriptions indicates an increase in revenue and a user engagement success. On the other hand, the number of users on a new platform doesn't mean they are active users, so no proven indicator of success is found on that number. In the process of analysis, additional categories than the ones proposed by García-Avilés et al. (2018) emerged. Also, others had to be adapted in order to better fit with the analyzed cases and one (prestige and impact) was retracted, due to its difficulty in measurement. All proven outcomes were analyzed considering its impact in journalism—not the innovation itself.

---

[2] All data refers to those presented in the Google DNI Fund three-year report (2018). Other sources of each initiative were not consulted, in order to maintain a pattern.

Finally, the fifth axis—*Potential replicability degree*—can assume a three-stage scale: (a) high replicability, when resources or efforts to experiment this innovation in other context were simple to be done; (b) medium replicability, when a considerable investment of resources or considerable process modifications, but not both, was needed; (c) low replicability degree, investment in both previous aspects was needed.

It is important to mention that the projects were previous categorized by Google into four clusters (Table 1) and for a better comprehension of each project, this division did not considered the main focus of each innovation, as the six axes discussed have proven to be a more comprised and detailed view of the innovations in question.

Table 1 Projects

Category (Google DNI Fund)	Project name	Country	Fund awarded (year)
Battling misinformation	*Full Fact*	UK	2016
	TrustServista	Romania	2016
	Factmata	UK	2016
	VIS Media	Italy	2017
	Full Fact	UK	2017
Boosting digital revenues	*REMP*	Slovakia	2015
	Steady	Germany	2016
	La Numérique	France	2016
	Nonio	Portugal	2017
	REMP	Slovakia	2017
	Gudbrandsdølen Dagningen	Norway	2017
Exploring new technologies	*Frames*	Portugal	2016
	Il Secolo XIX	Italy	2016
	The Buzzard	Germany	2017
	QuoteBot	Belgium	2018
Telling local stories	*Local News Engine*	UK	2016
	Tagesspiegel LEUTE	Germany	2016
	The Bureau Local	UK	2016
	La Voz de Galicia	Spain	2016

Own elaboration

5 Discussion: What Do Google DNI Fund's Choices Tell Us About Innovation in Journalism?

Our research is an exploratory work with a mixed approach (quantitative and qualitative): due the small dimension of the sample (seventeen projects), we emphasize the qualitative approach based on the published interviews with those responsible for the projects supported by Google.

5.1 Implementation Stage

Most of the initiatives (thirteen) were already public, that means, in the adoption level. One was still in development, two were a working prototype and the other was in internal testing phase by the time the report was released.

The high percentage of projects at the adoption level proves a trend throughout the history of journalism: companies test new tools or procedures before the academic research has results, which is the opposite of what happens in other economic sectors where products reach the market after tests carried out at universities and research institutes.

5.2 Innovation Mindset

Collaboration or wider impact is pursued in most cases: more than half of the projects (nine out of seventeen) held actions towards an ecosystem impact. Besides them, three more demonstrate an ecosystem view or ecosystem plan within their strategies. It indicates that even if competition is high in the news industry and an important reason to innovate (García-Avilés et al. 2018), innovation is also tied with a broader survival goal and some kind of network approach, support, collaboration or industry-wide target. Examples of that are the initiatives that were born with an open source strategy as *Local News Engine*, in the United Kingdom, or *REMP* (*Readers' Engagement and Monetization Platform*), in Slovakia. There are also others as *Full Fact*, which states as a key goal "sharing of experience, expertise and tools across the fact-checking community" (Google 2018: 9). More than just a statement, they are already working together with similar organizations in Argentina and South Africa, even though they are based in the United Kingdom.

The innovations analyzed come from a wide set of institutions: charity and nonprofitable institutions, new start-ups and legacy media business. Open source is not a viable or interesting strategy for all of them (in fact, those two are the only ones with this specific mindset), but that does not mean the spread or ecosystem impact of the others is set aside. Some of them, such as *TrustServista*, in Romania, are targeting an industry-wide impact using a software-as-a-service strategy. In their cases, it is also a sustainability road. Ecosystem thinking is not only related to non-profitable collaborations or interactions but, in fact, connected to a wider innovation impact, whether profitable or not. Besides those already mentioned, others, as *Nonio*, in Portugal, have their own history based on collaboration: this one is the result of a consortium between media companies, starting from scratch with an ecosystem impact mindset, even if their impact still needs to be proven.

5.3 Innovation Focus/Innovation Target

There is an interesting point regarding the innovation targets. Seven of the seventeen projects are directly connected with new business models or impact in journalism sustainability, precisely one of the goals of the *Google DNI Fund*. They are doing it in much diversified ways. For instance, *Facmata* is using artificial intelligence "to assess the risk to advertisers of appearing alongside what could be inappropriate content" (Google 2018: 12). *Steady* provides "journalists with the technology and marketing tools they need to build a membership base and generate revenue" (Google 2018: 24). However, most of them are not proposing a change on the traditional revenue streams, most common in the news industry (advertising and subscriptions). Even *Nonio*, which results from a collaboration between six Portuguese media companies, is still seeking to monetize journalism by a "compelling digital proposition for advertisers" (Google 2018: 26). One exception, though, is *VIS Media*, which is pursuing a software as a service strategy for its machine learning fact-checking solution, having a combined innovation focus in new journalism production or processes and new business models or impact in journalism sustainability.

News content (narratives, formats or niche content) are the focus of six projects and we know that the news language and/or new user segmentation is a potential variable to attract readers and, consequently, more revenue. In fact, all six projects related to this innovation target mention its search for increasing or diversifying revenues as a desired correlated consequence,[3] most of them connected with local communities or new niches. One example is the German initiative *Tagesspiegel LEUTE*, which highlights in its description a 40% rise in subscribers, resulted of providing local news to readers in twelve districts of Berlin. It is evident that they are innovating in news content (narratives, formats or niche content) tied with a complementary medium or long-term desire to increase profits, mostly through the same old and traditional models, as, for instance, subscriptions. The exceptions of this kind of relation with business profits are *Il Secolo XIX* and *The Buzzard*: the former is an online social media-training platform targeting in-house journalists to "create compelling stories that pull readers into the website" (Google 2018: 36) and the latter is a "machine

---

[3] A remark: projects can be connected with more than one innovation target, but it should be a primary focus. In case of news business models it has, for instance, to be related with a paid product or a solution directly connected with this profits. If it is a secondary goal resulting from a long-term strategy, as new niche audience to have, in the future, more subscribers, it was not considered as a new business model primary target.

learning algorithm to include all sides of a debate and provide users with a balanced, holistic overview" (Google 2018: 34).

A third group of projects is connected with social innovation (five projects). Those are related with fact-checking tools (as *Full Fact*), trustworthiness of news stories (as *TrustServista*) or empower hyperlocal news through a user centered perspective and social mindset (*Local News Engine, La Voz de Galicia, The Bureau Local*). Those were considered social as they have mentioned a social related concern in their description.

The journalism production or processes have the same number (five projects) as social. It ranges from fact-checking and/or trustworthiness automatization (*Full Fact, TrustServista, VIS Media*) to news writing or lead generation automatization (*QuoteBot, Local News Engine*). This number shows that the field might be finally considering to target on processes, rather than just on products.

Channels of news distribution had only two initiatives: it was *La Numérique* which has created a new digital evening product and *La Voz Galicia*, which has created a platform that indicates the most appropriate media mix to publish and promote the stories created in newsroom, being so, an innovation in news distribution.

5.4 Proven Journalist Outcomes

Although willing to have an industry-wide impact, seven of the seventeen projects did not mention any data that could indicate an effective consequence for journalism. *Nonio* is one example. There are two numbers highlighted in the report related to them: one is reachability in the local market and the other is the total amount of users in the platform. Reachability in local market does not affect journalism itself, even if it can be a way for the platform to be relevant in the future. The number of registered users does not mean engagement, as we do not know how many are still using the product and with what frequency. *Il Secolo XIX*, an Italian in-house digital media-training platform for journalists, is another example. Even if one in six journalists from the company have completed the course so far, there is no evident consequence or numbers highlighting growth in audience, revenue or other journalistic outcome for its holding media company.

The other ten projects were able to present numbers related to effective journalistic impacts or innovation success:

(a) one evidence of audience growth: *Frames* (Portugal) indicated a 5.6% increase in pageviews in *Observador* website;
(b) five evidences of user satisfaction and/or engagement with journalism as well as five of increasing revenues, two topics strongly connected; as an example, we can cite the 98% of subscribers retained monthly in a new platform called *Steady* (Germany) and an increase in the subscriber base to over 120,000 in *L'Equipe* (France) through *La Numérique* innovation;

(c) four has shown indicators of improvement in processes, as the 30 min saved per journalist per day by using *QuoteBot*, from Belgian financial publisher *MediaFin*, or the 60,000 website articles fact-checked per day by *TrustServista*.[4]

One topic should be pointed out in relation with proven outcomes: regarding those directly targeting social impact (related with society benefits and possible with journalism's democratic goal), none of them presents numbers capable to prove its influence in this sphere.

5.5 *Potential Replicability Degree*

The majority of projects (nine) have a high replicability degree, followed by seven projects with medium replicability potential. This fact is also true in projects presenting a unitary view, which is an interesting fact: even having a medium or high replicability potential, some initiatives are still thought with an internal perspective, mainly focused on solving contextual companies' problems instead of having a broader perspective on solving the challenges of journalism as a whole. Many of them could be easily replicable as a software as a service or an open source strategy, for instance. Others, however, are tied with people investment and/or local data particularities.

6 Conclusion

Innovation is a key element in any industry, having a central place in the media ecosystem because of the accelerated digitization in progress since the end of the 20th century. To keep up with the innovation cadence, the media has chosen mainly one of four paths: building up an internal innovation team, launching its own research laboratory, outsourcing research or developing stand-alone solutions to specific problems. In any of these solutions, financial resources are critical, so companies, laboratories and other organizations seek support, and the *Google DNI Fund* emerge as one of the financing alternatives. To study the media innovation, in this research we are analyzing seventeen projects financed by this fund in the last three years and to carry out this analysis, a grid was developed based on previous researches, but with adaptations.

Most of the solutions are already in the process of adoption, but they were expected to arrive on the market only after months or years of development and testing. This trend is particularly noticeable in projects with impacts on the ecosystem, which confirms that the innovation is complex and always a long-term process, as mentioned by Ludovic Blecher, head of the Digital News "In the early days, we saw a lot of brilliant ideas begin their journeys. But execution takes time and now we're at the

---

[4]Each innovation initiative could be connected to more than one proven journalism outcome.

stage where some of these ideas start having an impact where it matters most: in newsrooms" (Google 2018: 3).

Although media is a very competitive sector, there is a trend towards ecosystem-oriented solutions. In fact, rather than delivering innovative but single application solutions, the media need universal solutions because users can only decode the message if they master a set of intellectual tools. Just as language mastery allows us to read a text, media literacy is essential to decode news content in terms of language, organizational structure, etc. The proposal does not target media literacy directly, but they endorse the same long-term or broader mindset comprised in this thought. That's why all the proposals, in some way, are contributing to journalism's subsistence.

News business models and news narratives or formats are the majority of the projects innovation goals and they are strongly connected, which highlights a trend that also appears in the next points: the need to respond to readers' wishes. In fact, it is remarkable how most of them are connected directly or complementary with the search for rising journalism profits. It is interesting, though, that most of them are still targeting the same traditional revenue models: subscriptions and advertising. The third goal is social, and it turns around the fact-checking projects and empowerment of local news. Although being seen as a particular characteristic on innovation in journalism, it is remarkable the inexistence of proven outcomes: are they impacting the public view of journalism? Is this effort helping to decrease fake news or misinformation? If social goal is a major particularity of journalism innovation, the industry should develop better ways to assess and measure its innovation impacts. It might not be easy, but necessary.

Although none of the projects have the innovation focus in the community engagement, this is supposed to be one of the major goals of journalism nowadays: how to better engage readers in journalism. In our perspective, it could be because the media consider that technologies already exist to bring the newspapers closer to the readers, they just are not being used correctly. It could also be connected to a mindset that journalism should still be primarily led by journalists, rather than a shared process between professionals and the audiences. However, more research is needed in order to further understand this topic.

In terms of proven outcomes of journalism, we note a concern with the recovery of proximity to readers and their satisfaction through news content innovation focus. In this field there is a noticeable distance from the old journalism that advocated 'we write, you read'. In this goal it is mandatory to highlight the strong connections between the user satisfaction/engagement with journalism and the increasing of revenues, which proves that the media have already realized that it is not possible to survive without meeting readers' expectations.

Also noteworthy are the concerns about the production processes, finally realizing that, in a way to have multimedia journalists, it is necessary to find tools that facilitate some of the traditional steps of the production process, such as data collection. In another view, also noting that journalism production might have an opportunity to overcome the phase of just using technology, to build up new tools and products from scratch to the improvement of journalism processes.

Finally, it should be noted that the highly replicable projects are the most represented, confirming the situation described above. The crisis forced the media to think in an ecosystem perspective rather than a demand for individual solutions, confirming a tendency towards open and autonomous systems over closed owner solutions. A question remains, however: are other sources of journalism innovation around the world, also opting for an ecosystem view? Even though it is a noticeable fact concerning the analyzed initiatives, it is still uncertain if the news industry as a whole or its majority has already realized that in order to grow they need to focus on a coopetition (Bengtsson and Kock 2000) rather than a simple competition. Journalism survival might be depending on that.

References

Bengtsson M, Kock S (2000) "Coopetition" in business networks—to cooperate and compete simultaneously. Ind Mark Manage 29(5):411–426

Bower JL, Christensen CM (1995) Disruptive technologies: catching the wave. Harv Bus Rev, 43–53

Christensen G (2016) Panel on reproducible journalism. Berkeley initiative for transparency in the social sciences. Retrieved from https://www.bitss.org/2016/10/26/panel-on-reproducible-journalism/. Accessed 17 Apr 2019

Christensen CM, Skok D, Allworth J (2012) Breaking news—mastering the art of disruptive innovation in journalism. Nieman Rep 66(3):6

Dogruel L (2013) Opening the black box. The conceptualising of media innovation. In: Storsul T, Krumsvik AH (eds) Media innovations: a multidisciplinary study of change. Göteborg, Nordicom, pp 29–43

Ferguson DA (1991) The domain of inquiry for media management researchers. In: Broadcast education association annual meeting, conference proceedings, Las Vegas, pp 1–16. Retrieved from: https://www.researchgate.net/publication/264120088_The_domain_of_inquiry_for_media_management_researchers. Accessed 20 Apr 2019

García-Avilés JA, Carvajal-Prieto M, Arias F, De Lara-González A (2018) How journalists innovate in the newsroom. Proposing a model of the diffusion of innovations in media outlets. J Media Innov 5(1):1–16

Google (2018) Elevating quality journalism: digital news innovation fund report 2018. Retrieved from https://newsinitiative.withgoogle.com/dnifund/report/. Accessed 10 Jan 2019

Jenkins H, Ford S, Green J (2018) Spreadable media: creating value and meaning in a networked culture. New York University Press, New York City

Kovach B, Rosenstiel T (2014) The elements of journalism: what newspeople should know and the public should expect. Three Rivers Press, CA

Küng L (2013) Innovation, technology and organisational change: legacy media's big challenges. In: Storsul T, Krumsvik AH (eds) Media innovations: a multidisciplinary study of change. Nordicom, Göteborg, pp 9–12

Lavine JM, Wackman DB (1988) Managing media organizations: effective leadership of the media. Longman Pub Group, New York

Lindmark S, Ranaivoson H, Donders K, Ballon P (2013) Innovation in small regions' media sectors. In: Storsul T, Krumsvik AH (eds) Media innovations: a multidisciplinary study of change. Nordicom, Göteborg, pp 127–144

OECD (2005) Oslo manual: guidelines for collecting and interpreting innovation data. In: Organisation for economic co-operation and development. Statistical Office of the European Communities,

Paris, p 163. Retrieved from https://ec.europa.eu/eurostat/documents/3859598/5889925/OSLO-EN.PDF/60a5a2f5-577a-4091-9e09-9fa9e741dcf1. Accessed 10 Mar 2019

Posetti J (2018) Time to step away from the 'bright, shiny things'? Towards a sustainable model of journalism innovation in an era of perpetual change. Journalism innovation project. Reuters Institute for the Study of Journalism, Oxford University, Oxford

Storsul T, Krumsvik AH (2013) What is media innovation? In: Storsul T, Krumsvik AH (eds) Media innovations: a multidisciplinary study of change. Nordicom, Göteborg, pp 13–26

Ana Cecília B. Nunes Assistant Professor at Pontifical Catholic University of Rio Grande do Sul–PUCRS (Brazil) and academic head of IDEAR, an interdisciplinary entrepreneurship and innovation lab at PUCRS. Currently, she is Ph.D. candidate in a joint degree from PUCRS University (Brazil) and the University of Beira Interior (Portugal)—Capes/PDSE fellow. Her research focuses on challenges of media innovation, particularly in the field of journalism. She is also part of Ubilab, a digital media lab at PUCRS.

João Canavilhas Associate Professor at Universidade da Beira Interior (Covilhã, Portugal), where he actually is vice-rector and researcher at Labcom.IFP—Communication, Philosophy and Humanities. His research work focuses on various aspects of Communication and New Technologies, particularly in the fields of online journalism, e-politics, social media and journalism for portable devices.

Innovating Journalism by Going Back in Time? The Curious Case of Newsletters as a News Source in Belgium

Jonathan Hendrickx, Karen Donders and Ike Picone

Abstract Among the slew of innovations piercing legacy news media in order to maintain their importance, relevance and financial viability, one notable shift in online newsrooms is the resurgence of newsletters as a controlled means of disseminating curated news content and controlling incoming traffic on news websites. This is particularly the case in Belgium, where nearly a quarter of the population indicated newsletters as their primary source of news in the 2018 Oxford Digital News Report. In this chapter, we establish three main reasons explaining the sudden rebirth of newsletters, and zoom in on one leading Belgian media player to show that, how and why (a) newsletters have emerged as bigger sources of incoming online traffic than social media and (b) newsletters have effectively altered the daily work of journalists in online newsrooms, as newsletters have become a focal point in the 'digital first' approach adopted by legacy media.

Keywords Digital journalism · Newsletters · Digital first · Innovative journalism

1 Introduction

In recent years, the production of e-mail newsletters has sharply increased, both by 'legacy' print and newer digital media publishers. The trend mirrors the continued strength of e-mails in daily life, and their widespread use in marketing, despite the advent of more sophisticated and proprietary digital tools (Jack 2016). Written off entirely a few years ago, and still threatened by social media and chat applications such as Facebook and Slack, e-mails in general and newsletters in particular have somehow managed to not merely survive, but even thrive in the era of endless supply

J. Hendrickx (✉) · K. Donders · I. Picone
Vrije Universiteit Brussel, Brussels, Belgium
e-mail: jonathan.hendrickx@vub.be

K. Donders
e-mail: karen.donders@vub.be

I. Picone
e-mail: ike.picone@vub.be

© Springer Nature Switzerland AG 2020
J. Vázquez-Herrero et al. (eds.), *Journalistic Metamorphosis*,
Studies in Big Data 70, https://doi.org/10.1007/978-3-030-36315-4_5

of both news and news sources. Zooming out, the renewed adoption of newsletters in newsrooms is also a compelling example of how certain media forms deemed obsolete at one point can maintain or regain relevance when it ticks all the right boxes.

The small, Western country Belgium always makes for interesting case studies in media research due to its convoluted linguistic and political situations, effectively yielding multiple media markets within one country (Donders et al. 2019). It is therefore the more surprising that Belgians appear to be united when it comes to consuming news via e-mail newsletters. According to data from the 2019 Oxford Digital News Report (hereafter 'DNR'),[1] e-mail newsletter or notifications are used by 30% of Belgian news users to access news on a weekly basis. This however hides a significant difference between Dutch-language northern region Flanders and French-language southern region Wallonia: 38% and 23% respectively. Even if the popularity of newsletters is slightly declining over the past years (down from 40% in 2016), it remains a key access point to news in 2019 in Belgium. Especially in Flanders, it forms a more popular way to access the news than direct access (34%), social media (26%) and mobile news alerts (16%). A quarter of Flemish news users even consider newsletters and notifications via e-mail their main way of accessing news.

Again, these numbers conceal important differences amongst Flemish news users. For 30% of news users above 35 years old, newsletters and notifications via mail form their main way of getting the news. This drops to only 8% for those below 35. A similar gap can be noticed in terms of people's education: around a quarter of low- and middle-educated Flemish news users report newsletters to be their main way of accessing the news, while this is only 10% among the high-educated.

These important nuances can give us a hint as how to explain the popularity of newsletters in Flanders. The prominence amongst older news users might point towards the sustained centrality of mail as a means of communication within older age groups. Older news users might have gotten accustomed to using newsletters several years ago, developing into persisting habits. The younger generations however already seem to connect less with this form of news. The higher adoption across low- and middle-educated news users might be due to the fact that over recent years especially more popular news brands have invested in their newsletter offering. For example, newsletters with the latest updates from one's town or village formed a key component of the renewed strategy to focus on local news a few years ago of the Flemish newspaper *Het Nieuwsblad*'s, one of the four papers this chapter will further discuss and analyse. Nonetheless, in no other country included in the 2019 DNR is e-mail so popular as in Belgium, with considerably lower popularity in key media markets such as the US (21%), the UK (10%), France (17%) and Finland (9%) (Nielsen et al. 2019).

In this book chapter, we aim to contribute to the thus far very limited amount of available literature on newsletters in the 21st century. We do this by pinpointing three

---

[1] As the Belgian partner in the Digital News Report Consortium, we have accessed the primary data to provide these numbers.

reasons explaining why editors are including newsletters in their offering, which will be further elaborated upon in the literature review and analysis parts. We focus on (a) dependence on social media's algorithms and the power of platforms in regulating web traffic; (b) attempts to regain customer ownership and returning to the gatekeeping and agenda setting-functions of journalism and (c) the diversification of news offers to increase the overall reached online audience. Our findings are based on literature, ethnographic observations, expert interviews and traffic data analyses of the Belgian popular newspaper *Het Nieuwsblad*, part of the Mediahuis media corporation which owns newspapers, radio stations and classifieds in Belgium (Flanders), the Netherlands and, since 2019, Ireland. Through *Het Nieuwsblad* and its pivotal role for newsletters in news dissemination and media innovation strategies, we are able to construct the story of why and how this once considered antiquated means of communication has against all odds become in vogue again.

2   Newsletters' Rise, Fall, and Rise

The body of literature specifically zooming in on the resurgence of newsletters among news media is as of yet peculiarly small and in desperate need for expansion, particularly because of the widespread popularity of newsletters throughout both legacy and online-only media. After a series of intense searches on Google Scholar and in Web of Science, we found only a small handful of recent publications and reports scratching the surface of newsletters as a verified news platform in the digital age of journalism (see, among others, Jack 2016; Santos and Peixinho 2017). This book chapter therefore fills an existing gap in academic research into this small yet indispensable part of media companies' digital strategies.

The newsletter is older than one would think and was the precursor of regular newspapers, and predates the term journalism by about two centuries as that term was introduced in the English language only in the 1830s. Newsletters had first emerged in the early 15th century as written and printed translations of foreign events, natural disasters and supernatural occurrences (Rubery 2010). The notion of providing news about the contemporary world at regular intervals eventually evolved into the current multiplatform mass-media age. Interestingly, while being roughly 400 years old, newsletters have far from disappeared as a useful means of disseminating curated news content to audiences.

Just a few years ago, funeral services were held around the world to mourn the demise of e-mails. The advent of instant messaging and social media apps would logically lead to the end of e-mails, which had been considered an old-fashioned technology from the early days of the Internet, and until recently most experts seemed to agree that its death was nigh (Fagerlund 2016).

> I thought newsletters were an outdated technology, something for old people. The Washington Post was quite popular on Facebook, but suddenly they changed their algorithms and we lost a lot of readers. I realized that I needed to find something where I can control the means of

production. Newsletters were one way of doing it (David Beard, director of digital content of The Washington Post in Fagerlund (2016).

In a report for the LSE's media think-tank Polis, Fagerlund (2016) describes the change in attitudes towards e-mail newsletters, claiming that it was initiated by individual American journalists who decided to reach their audiences directly through their inboxes, exactly as early adopters of the 1990s had previously already done through 'Listservs' (Jack 2016). Companies like Mailchimp facilitate sending out e-mail newsletters and have replaced blogs as the main choice to reach (new) audiences. Online-only media such as *The Skimm*, *Quartz* and *Buzzfeed* launched their own e-mail newsletters around 2014, swiftly followed by major legacy print media such as *The Wall Street Journal* and *The Financial Times* (Fagerlund 2016). The trend then rapidly spread to other countries around the world, in a similar vein by first reaching online-only media and only afterwards established media outlets. While gathering data for their case study of how a new Portuguese online newspaper nationally pioneered and applied daily newsletters as a tactic to reach audiences, Santos and Peixinho (2017) found that legacy newspapers in the country too had started sending out several newsletters a day, often at a set hour.

While newsletters are thus the predecessors of newspapers, it is noteworthy that a study by Fredriksson and Johansson (2014) specifically found journalists who worked for organisations producing newsletters to be more often female than male, and more working as freelance journalists. They also found that this group of journalists is typically less embracing of the traditional journalistic ideals and tends to promote the amusement function of journalism.

3   Newsletters' Resurgence Explained

As discussed in the introduction, we pinpoint three key reasons explaining the rationale behind media's increased newsletter output. These will each be discussed separately underneath. We will take a closer look at the power of platforms, regaining customer ownership and the diversification of news offers.

In just over a decade, social media have effectively changed the lives of a few billion people worldwide and the way in which they receive information and consume media content online. News media quickly adopted to the sudden monopoly of particularly Facebook in acquiring vast amounts of online attention time and engagements by becoming active on them and posting their own content, in order to lure social media users to news media's websites. But a few infamous changes in Facebook's algorithm have turned social media into necessary evils for both legacy and online-only news media. Sizeable parts of populations remain active on social media on a daily basis, so it remains vital to maintain a strong and updated presence to gain readership—but Facebook is keen on keeping its users inside its own network. The renaissance of newsletters has thus been realised not in spite, but because of social media. It is a means to direct traffic to news websites and to ensure consumers are not locked into ecosystems of Google, Amazon and Facebook. Indeed, these platforms integrate both intermediation and gatekeeping functions. In combination with their economies of scale and scope, and diversification of activities, the ultimate aim is to construct an online world where users move from one website to another, and from one service to another, without actually leaving the platform's universum. This already happens from time to time, without consumers realising it (Moore and Tambini 2018) and was recognized as a key concern by several news media CEO's in both Norway and Belgium (Donders et al. 2018).

The move from a media environment dominated by direct discovery to one increasingly characterized by distributed discovery has forced news organisations to succeed their privileged and dominant position of gatekeepers to the likes of Facebook and Google (Kalogeropoulos et al. 2019). The dependence of news outlets on social media has become a thorny issue and has led to a myriad of attempts to regain control over the customer relationship. Email newsletters can form a key mechanism in these attempts, as they offer news organisations a direct and unfiltered access to a place where many Internet users reside and still spend a lot of time: their mailbox. This feature makes them a key component of news media's conversion funnel: they form an attractive entry point for new readers, turn casual readers into engaged ones by offering curated content, convert engaged readers into paying customers and keep loyal subscribers interested to reduce churn (Boltik et al. 2017). However, in today's news economy, "the supply of public attention is limited, and, since the endless number of claimants, scarce" (Webster and Ksiazek 2012).

By turning to newsletters, editors try to retain public attention by harnessing the power of habits, which they once mastered to perfection. News habits have always been an important aspect of news consumption. News users tend to return to their favourite news sources throughout the day "to relieve their vague sense of unease about not knowing what is 'going on' in the world" (Hartley 2018: 7, building on Diddi and LaRose 2006). Here too, news media face tough competition, again in first instance by social media who with their features such as feeds, 'likes', comments, tags, etc. seem designed to get users 'hooked' (Andreassen 2015: 179). Still, email newsletters are one of the most reliable digital channels editors have to their disposition to build a 'habit of news'. Not only are they delivered in mailboxes that are still central to many people's Internet use, but they also allow publishers to maintain the customer relationship and, hence, to collect user data to build behavioural models to maximize reader attention (Boltik and Mele 2017).

It becomes apparent that newsletter have the potential to help publishers regain (some) control over the customer relationship and find new revenue streams, but they are not a magic bullet. In order for newsletter to really convert casual readers to paying subscribers, they need to be more than a simple collection of links. This in turn requires newsletter to be given the necessary editorial attention, but on top of that also synchronized production with the marketing team, testing, and analytical work (Hansen and Watkins 2019).

Ultimately, the goal of retaining control over the flow of incoming traffic to online platforms of news outlets is to enhance overall readership and revenue, which subsequently leads to increased advertisement revenues. But journalism is no longer a one-size-fits-all-solution in which one newspaper is printed for the entire population. Personalised news is clearly here to stay, and newsletters have the opportunity to play a key part in this approach. Newsletters can be tailor-made to cater to very niche groups and communities, ranging from age groups, cities, towns and neighbourhoods to interests and types of news content. Specific newsletters have millennials, women, businessmen and-women, football enthusiasts and fans of lifestyle and/or fashion as target audiences. They can be a complimentary service offered to subscribers by providing overviews of articles behind paywalls, can contain coupons, discounts and contests and can be curated by individual editors of journalists offering handpicked selections of articles, just as the original online newsletters of the 1990s and early 2010s. This allows news outlets to diversify their output to the people who will most likely read it, bringing their journalism closer to its intended core audience.

4   Newsletters and Mediahuis

Media users' readership of newsletters is expected to be surprisingly large in Flanders, as previously shown by Digital News Report figures. We therefore want to investigate if this also shows in the ratio of incoming online traffic for legacy Belgian news outlets and if and how this has impacted the daily work of journalists working in Belgian newsrooms. For our study, we rely on ethnographic observations and expert interviews inside the newsroom of *Het Nieuwsblad*, the second largest and a popular Flemish newspaper owned by the Flemish media conglomerate Mediahuis. Since its inception in 2013 after a successful merger of two previously separately functioning media companies, it also owns and publishes three of the other in total seven daily paying Flemish newspapers: quality newspaper *De Standaard* and regional newspapers *Gazet van Antwerpen* and *Het Belang van Limburg*. We perform an analysis on incoming traffic numbers of the websites of the four Mediahuis-owned papers between January and March 2015 and 2019 (going back further in time was not possible) as found in *Traffic*, the internal data platform of Mediahuis where all available user data and official figures regarding readership of online news are aggregated in a user-friendly platform intended for usage by the journalists working for the company. With the analysis, we intend to gauge the changes in incoming traffic sources and try to see what they indicate for Mediahuis, journalism in Flanders, and beyond.

4.1   Newsletters and 'Digital First'

Legacy newsrooms around the world are in varying stages of converting to a so-called 'digital first' approach, overtly placing news posted on its official website and app (and its social media accounts) as the key platform to post content as quickly as possible. Older media (newspapers, TV channels) remain the driving forces, but are no longer the number one priority in cases of breaking news; in such scenarios it is of course much easier for one online journalist to quickly publish an article of a few lines explaining the event and updating that article constantly than printing an extra newspaper or getting an extraordinary TV news broadcast started up technically.

The 'digital first' wave sweeping newsrooms not only externally but also internally wishes to put online news higher up the pecking order: journalists formally writing for newspapers are encouraged to also write for websites, and in many cases, they are more and more expected to become all-round journalists, providing self-created and -edited pieces for the TV and/or radio news and an article for the website on one given news topic. This is dramatically altering the daily work of journalists in newsrooms globally and occasionally proves problematic.

Mediahuis and its four newspapers have been no exception. Its main newspaper, *Het Nieuwsblad*, announced on its staff day in November 2017 that it had the ambition to become completely 'digital first' by 2020. After that announcement, it would take nearly 1.5 years for the first fully-fledged trial version to come into effect: in March 2019, the newsroom became more integrated. Journalists previously being the only ones to write articles for the website were divided in a so-called "in" and "out" team, with the former group still providing online news content and the latter becoming responsible for editing articles and disseminating them across social media and newsletters. Print journalists are expected to contribute to the website as well by providing unique news content throughout the day, with their output often published online behind the paywall and, sometimes in an abridged and/or updated form, republished in next day's print newspaper.

Perhaps the biggest change of the 'digital first'-tactic for *Het Nieuwsblad* has been the increase of deadlines and their specific focus on newsletters. Currently, four daily e-mail newsletters are sent out, each with their own content and focus, and three with their own deadlines. There are also weekly and other extra newsletters, but for the sake of brevity, we focus on the four daily ones here. Paying subscribers automatically receive all newsletters, while non-subscribers can opt into receive separate e-mails.

The morning e-mail is sent out around seven in the morning and focuses on the main stories of that given morning, both from *Het Nieuwsblad* and other media. The second e-mail is due just before noon to give readers the opportunity to read it during their lunch break, and has a slightly lighter tone. The third newsletter is usually sent out around 4:30 in the afternoon and contains the biggest stories of the day and a few more articles behind paywalls to lure readers into considering purchasing a subscription. Finally, the evening newsletter is sent out at 7:30, but is solely intended for subscribers of the newspapers, and focuses on five or six main articles for next morning's newspaper. For all newsletters, a number of articles is 'ordered' to be finished by then, so that they all have new and unique content to present to their readers. The decisions on the content are made predominantly by the news editors who decide what news to present where and when, instructing the online journalists specifically designated to catering all news dissemination across social media and newsletters, called the "out" team as they "push out" the news to media users, what to put where in which newsletter. This highlights the in-between nature of newsletters. They are both editorial and business products, and unlike an advertisement or a news story, their place within news organizations is an open question. One could argue this makes newsletters a convincing mirror of the state of the industry today (Porter 2018).

11:45 h, 16:30 h and 19:30 h have effectively become deadlines throughout the news day at *Het Nieuwsblad*, at which every day a few articles need to be finished to be sent out as key articles in the e-mails. The main deadline is still 22:30 h, which is when next day's newspaper is sent to the printer and needs to be completely finished, but the advent of 'digital first' and the increased importance given to e-mail newsletters have thus considerably changed the work of journalists as for many of them, the only deadline they had was 22:30 h. They are now expected to provide finished articles at noon or in the afternoon, greatly impacting their daily routine.

4.2 Newsletters in the Mediahuis Data

Have these changes in the daily proceedings of journalists in order to make newsletters more present in their digital strategy and more topical and urgent towards news users had any effect on readership of e-mail newsletters of said users? When looking at the available data on incoming web traffic at the internal platform of Mediahuis, *Traffic*, we come to an interesting finding which goes against the trend of increased e-mail newsletter readership: the percentage of visits to the four Mediahuis news websites through newsletters actually decreased over time. In the table underneath (Fig. 1), we summarise figures applicable for the first three months of each given year, in weighed percentages for the four newspapers together. Traffic makes a logical distinction between direct traffic (e.g. people going straight to the website or app of a newspaper), social traffic (through social media accounts—this includes so-called "dark social" traffic which is impossible to trace back to its source and is likely shared

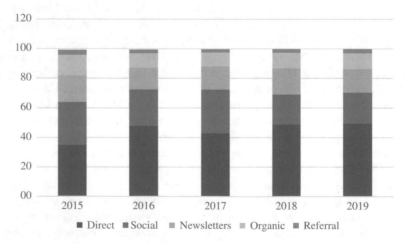

Fig. 1 Incoming traffic sources for Mediahuis websites (2015–2019). Internal Mediahuis data platform

through chat applications such as WhatsApp which provide end-to-end encryption), newsletter traffic, organic traffic (through unpaid search results at engines such as Google) and referral traffic (through clicking links on other websites).

The curve for newsletter traffic is rather capricious, with 17.9% for early January to late March 2015 to 14.7% for 2016, 15.9% for 2017, 17.7% for 2018 and finally 16.0% for the first quarter of 2019. The changes in the Facebook algorithm are clearly reflected as well, as the percentage of social traffic plummeted from 29.0% in 2015 to just 20.9% four years onwards. All this is somewhat saved by the vast increase in direct traffic, amounting for 35.0% in 2015 and nearly half of all incoming traffic (49.5%) in 2019. Thus, even though there are of course other assessment factors in order to gauge the success rate of e-mail newsletters, it appears from the raw user data presented here that the popularity of newsletters among Belgian news consumers has not (yet?) created many benefits or enhanced online traffic for the four Mediahuis news websites.

It is also noteworthy that there are differences between the four different websites (Fig. 2). Leading newspaper *Het Nieuwsblad* had the biggest increase in newsletter traffic (17.9% in 2015 to 21.5% in 2018), but then saw it decrease again to 18.0% for the first three months of 2019 (overall increase of 0.1%). The other three Mediahuis papers saw bigger slumps for its newsletter traffic percentages over four years' time: *De Standaard* dropped from 15.4% to 12.3%, *Het Belang van Limburg* 19.5–12.8% and *Gazet van Antwerpen* from 20.2% to 15.0%, respectively signifying 20.3, 34.5 and 25.2% drops in the overall share of incoming traffic sources.

For all newspapers, the share of direct traffic rose considerably over our given time frame, in line with the numbers of the DNR. Interestingly, popular newspaper *Het Nieuwsblad* is the poorest performer here with a 44.6% share for direct traffic in 2019, whereas its three sister newspapers all exceed the 50%-mark with ease. This finding

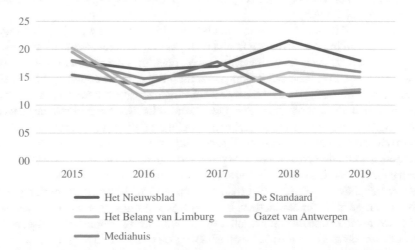

Fig. 2 Share of incoming newsletter traffic on Mediahuis websites (2015–2019). Internal Mediahuis data platform

is to be explained by the increased popularity of applications for mobile phones and tablets, which allow users to enter the online version of the newspapers directly. We cannot disclose detailed figures about this matter, but can confirm that daily usage of the apps of the four Mediahuis newspapers has indeed increased substantially over four years' time, yielding much more incoming direct traffic.

5 Conclusions

This chapter has discussed the unexpected increased popularity of e-mail newsletters as a successful means for legacy and new news media outlets alike to disseminate their content independently, without having to rely on algorithms of major social media. We have shown how a leading Flemish media company, Mediahuis, has effectively applied the newsletters as an integral part of its 'digital first' approach, and how this in its turn has altered the daily work of journalists working for its newspapers as new deadlines for articles have been added throughout a given working day, deliberately matching with the fixed points in time in which new newsletters are being sent out. An analysis of internal and incoming traffic data of the four Mediahuis newspapers has revealed that in spite of the attempts to put more focus on newsletters, this has not translated in higher shares for them in the overall incoming online traffic. As a matter of fact, the total share for newsletters has increased between 2015 and 2019, albeit not as sharply as has the share for incoming traffic via social media, which confirms the negative effects of Facebook's algorithm changes enacted from 2017 onwards. The share for direct traffic, on the other hand, has seen vast increases, which is to be explained by the shift towards using mobile phone and tablet applications to enter online versions of newsletters directly rather than through a social medium, a search engine or a web browser.

Time will tell if e-mail newsletters manage to either become more popular among media users and influential among media makers, or if their renaissance has perhaps already passed its peak. Two lingering issues regarding newsletters arise and will undoubtedly prove to become challenges for their future in the coming years. Firstly, newsletters at Mediahuis, and at the overwhelming majority of news media, are at the moment distributed entirely free of charge and considered as a complimentary service that news organisations offer its users to attract larger online audiences to its content. It is to be expected that newsletters will pass a stage in which their monetisation will be questioned, as manpower is needed to create and sent out the newsletters. Particularly when their effects on incoming traffic sources are not to be overestimated, as this very chapter has shown, we gauge that media companies will soon start thinking about ways to charge users for newsletters, by for instance promising more unique content and tailor-made newspapers for smaller niche audiences currently un(der)served by the bulk newsletters meant for all—considering they are not already doing so at the time of writing.

The second issue endangering a healthy future for e-mail newsletters as a viable distribution means is fatigue. With many legacy and new news media sending multiple newsletters a day, we venture that media users will reach a saturation point after which they will start caring less about newsletters because of their sheer abundance. Just as there is no need to read the same news in four different newspapers, the same applies to newsletters; a key task of news outlets is to distinguish themselves from their competition, even within one media conglomerate such as Mediahuis, and present their content in unique ways which continues to capture the attention and imagination of its (potential) visitors. Here too, we propose focusing on personalised newsletter experiences with offered content destined to persuade more receivers of newsletters to start consuming news content.

References

Andreassen CS (2015) Online social network site addiction: a comprehensive review. Curr Addict Rep 2(2):175–184

Boltik J, Mele N (2017) Practice, research, & technology. In: Using data science tools for email audience analysis: a research guide. https://shorensteincenter.org/email-analysis-research-guide/. Accessed 8 July 2019

Boltik J, Mele N, Practice et al (2017) Using data science tools for email audience analysis: a research guide. Available at: https://shorensteincenter.org/email-analysis-research-guide/. Accessed 8 July 2019

Diddi A, LaRose R (2006) Getting Hooked on news: uses and gratifications and the formation of news habits among college students in an internet environment. J Broadcast Electron Media 50(2):193–210

Donders K, Enli G, Raats T, Syvertsen T (2018) Digitisation, internationalisation, and changing business models in local media markets: an analysis of commercial media's perceptions on challenges ahead. J Media Bus Stud 15(2):89–107

Donders K, Van den Bulck H, Raats T (2019) Public service media in a divided country: governance and functioning of public broadcasters in Belgium. In: Połońska E, Beckett C (eds) Public service broadcasting and media systems in troubled european democracies. Springer International Publishing, Cham, pp 89–107. http://link.springer.com/10.1007/978-3-030-02710-0_5. Accessed 8 July 2019

Fagerlund C (2016) Back to the future-email newsletters as a digital channel for journalism. Journalistfonden. http://www.lse.ac.uk/media@lse/Polis/documents/Back-to-the-future—Email-Newsletters-as-a-Digital-Channel-for-Journalism.pdf. Accessed 20 June 2019

Fredriksson M, Johansson B (2014) The dynamics of professional identity. Journalism Pract 8(5):585–595

Hansen E, Watkins EA (2019) News media needs to convince readers to open their wallets. Consolidation has not helped. Columbia Journalism Rev. https://www.cjr.org/tow_center/newsroom-consolidation-product.php. Accessed 8 July 2019

Hartley JM (2018) News audiences and news habits. In: Oxford research encyclopedia of communication. https://oxfordre.com/view/10.1093/acrefore/9780190228613.001.0001/acrefore-9780190228613-e-845. Accessed 8 July 2019

Jack A (2016) Editorial email newsletters: the medium is not the only message. https://ora.ox.ac.uk/objects/uuid:8248179f-83e1-4bb9-81d1-6197d77900f3. Accessed 19 June 2019

Kalogeropoulos A, Fletcher R, Nielsen RK (2019) News brand attribution in distributed environments: do people know where they get their news? New Media Soc 21(3):583–601

Moore M, Tambini D (2018) Digital dominance: the power of Google, Amazon, Facebook, and Apple. Oxford University Press, New York

Nielsen RK, Newman N, Fletcher R, Kalogeropolous A (2019) Reuters institute digital news report 2019. https://reutersinstitute.politics.ox.ac.uk/sites/default/files/2019-06/DNR_2019_FINAL_0. pdf. Accessed 19 June 2019

Porter C (2018) Making a newsletter: what happens before nonprofit newsrooms hit "send"? Medium.   https://medium.com/single-subject-news-project/survey-workflows-for-newsletters-vary-across-nonprofit-newsrooms-debfcb1fc6fa. Accessed 8 July 2019

Rubery M (2010) Journalism. In: The encyclopedia of the novel. https://onlinelibrary.wiley.com/doi/abs/10.1002/9781444337815.wbeotnj003. Accessed 19 June 2019

Santos CA, Peixinho AT (2017) Newsletters and the return of epistolarity in digital media. Digital Journalism 5(6):774–790

Webster JG, Ksiazek TB (2012) The dynamics of audience fragmentation: public attention in an age of digital media. J Commun 62(1):39–56

Jonathan Hendrickx Early-career Ph.D. Researcher, studying the effects of media mergers on supply diversity as part of the inter-university DIAMOND project at the communication sciences department of the Vrije Universiteit Brussel (Belgium), imec-SMIT.

Karen Donders Professor in Policy Analysis, European Media Markets, and Political Economy of Journalism at the communication sciences department of the Vrije Universiteit Brussel (Belgium). She is a senior researcher of imec-SMIT. She specializes in (European) media policy, media economics, competition policy and media, and public service media.

Ike Picone Professor in Journalism and Media Studies at the communication sciences department of the Vrije Universiteit Brussel (Belgium). He is a senior researcher of imec-SMIT and is also affiliated with the Data and Design research group.

New Formats for Local Journalism in the Era of Social Media and Big Data: From Transmedia to Storytelling

Andreu Casero-Ripollés, Silvia Marcos-García and Laura Alonso-Muñoz

Abstract Local journalism is immersed in a scenario of structural transformation that questions its future. In this context, it is fundamental to bring innovation derived from social media and big data into the sector. The goal of this chapter is to provide a typology and classification of the new formats linked to social media that can be used by local journalism in news production. Based on an exploratory analysis six broad categories have been identified: storytelling, interactivity, multimedia, image, streaming, and transmedia. Within these six categories twenty-one different formats are located. On the whole, they suppose a broad and varied range of options so that local journalism introduces innovation to fight for its survival in the digital scenario.

Keywords Journalism · Local journalism · Social media · New formats

1 Introduction: An Uncertain Scenario for Local Journalism

Local journalism is immersed in a scenario of structural transformation as a result of the emergence and consolidation of digital technologies. These are introducing new ways to produce, distribute and consume news (Hess and Waller 2016; Salaverría 2019). So, they are challenging both the journalistic routines and the business models. For a decade, local journalism has been undergoing a strong economic reconversion accentuated by the impact of the economic crisis. As a result, the bases of its classic business model have been significantly weakened, and as of yet, no valid alternatives have been found to resolve a situation that seriously threatens the newspaper industry

A. Casero-Ripollés (✉) · S. Marcos-García · L. Alonso-Muñoz
Universitat Jaume I, Castellón, Spain
e-mail: casero@uji.es

S. Marcos-García
e-mail: smarcos@uji.es

L. Alonso-Muñoz
e-mail: lalonso@uji.es

© Springer Nature Switzerland AG 2020
J. Vázquez-Herrero et al. (eds.), *Journalistic Metamorphosis*,
Studies in Big Data 70, https://doi.org/10.1007/978-3-030-36315-4_6

(Casero-Ripollés 2010; Picard 2014). In this scenario, local journalism faces an uncertain future that questions its existence (Nielsen 2015).

Aside from challenges, the digital environment also opens up new opportunities for local journalism (López-García 2008). The low entry and operating costs potentially allow the emergence of new local and even hyper-local media (Negreira-Rey et al. 2018). Social media permits the implementing of new mechanisms of relating and interacting with the public, new strategies for community building and new narrative formats for the news (Newman 2009; Campos-Freire et al. 2016). All this favors the introduction of innovation in this sector. In fact, journalists perceive technology as a key element for innovation (García-Avilés et al. 2018). Discovering and taking advantage of the opportunities and developing new products and services are two pillars of innovation that apply to journalism (Küng 2015). It is not so much about incorporating digital technology into the production of news but about using it creatively and efficiently to renew journalistic content (Jarvis 2014).

In this context, local journalism must face the challenge of producing new content without abandoning the civic issues linked to local communities (Firmstone 2016) to help people define and maintain neighborhood identities and a sense of community belonging (Kim and Ball-Rokeach 2006). To do this, new formats to provide added value to local journalism amidst this uncertainty are essential (Casero-Ripollés 2014). Many of these, in addition, are based on the growing importance big data is acquiring in the field of journalism (Lewis and Westlund 2015). The abundance of information and the ease of access to the data that social media provides allows the creation and implementation of useful new formats for local journalism such as interactive maps or the geo-location of news. As a result, big data can be converted into a key ally of local journalism when it becomes time to innovate.

Social media has its own distinctive languages and communication codes. Likewise, these have generated specific new forms to narrate, configure, and disseminate journalistic information. These are characterized by the strong protagonism, on the one hand, of the visual elements and video (Kalogeropoulos and Nielsen 2018) and, on the other, by the increasing introduction of transmedia content (Vázquez-Herrero et al. 2018) and of big data. Local journalists are faced with the need to adapt to this new scenario and learn about their possibilities to produce news. For this reason, developing a catalogue capable of systemizing what these new information formats fostered by social media are becomes essential.

The aim of this chapter is to provide a typology and a classification of new formats linked to social media that can be used by local journalism in news production. In addition, the aim is to determine, in a critical way, characteristics, strengths and weaknesses of each of the identified formats. With this, a complete catalogue of new informational formats linked to social media is provided that local media can use to apply journalistic innovation and reach new audiences to cope with the uncertain situation affecting the sector.

The methodology applied is based on an exploratory approach supported by the technique of content analysis and it focuses on the study of three significant aspects of the new information formats for local journalism in social media. In the first place, is the typology of new formats linked to social media currently applied or likely to be used in local journalism. Secondly, are the defining characteristics of each of these formats to know what their potential and limitations in information terms are. Finally, the applicability of these formats to different local news events is considered.

2 Towards a Classification of New Formats in Social Media for Local Journalism

The following is a proposal for a classification to catalogue the new formats that the media can currently adopt, especially local media, in their news production and distribution processes. It specifically presents six major categories: storytelling, interactivity, multimedia, image, streaming, and transmedia. These categories integrate twenty-one different formats grouped according to their common characteristics (Table 1).

Table 1 Classification of new informational formats for local media linked to social media

Category	News formats
Storytelling	– Tweet magazine
	– Short messages published by journalists
	– News summaries through WhatsApp or Instagram stories
	– News coverage via WhatsApp or Telegram – Teaser – Videos using Instagram IGTV
Interactivity	– Interactive reports
	– Chat meetings
	– Voice notes using WhatsApp
	– Surveys, questions and questionnaires through Instagram
Multimedia	– Media reports
	– The expander
	– Atmosphere
	– Facebook instant articles
	– Podcasts
Image	– Infographics
	– 360° image and video
Streaming	– Live broadcasts through social media
	– Liveblogging
Transmedia	– Transmedia reporting
	– Collaborative reporting

Own elaboration

3   Storytelling

The concept of storytelling is understood as a set of techniques and strategies used to tell and share a story with the goal of creating added value to the news and generating public interest. It allows the use and combination of any type of language and format of the digital environment (text, image, video, audio, graphics, etc.) to build a story with a full narrative structure that is attractive to the audience. Thereby, information, adapted to the different tastes and demands of the audience can be presented truthfully without renouncing creativity.

In journalism, therefore, this technique is applied to any context that involves explaining the news in a way that differs from the conventional to be more attractive to the audience.

3.1   Tweet Magazine

It is a format built from the collection of tweets published by different users, usually the public, on their respective Twitter accounts. The journalist assumes the content curator role, since their function is to search, group and share what the most relevant published content has been on a topic or event in this platform.

One of the most significant advantages of this format is the ability to gather different views in the same space, favoring the contrast of news sources. The journalist can group tweets with different points of view to offer the user a comprehensive overview of the issue and generate debate. The overabundance of news is an obstacle, since the same subject can generate thousands of Twitter messages, so it is necessary to make a journalistic choice. Likewise, in the digital environment, where fake profiles and false news predominate, this requires the task of thoroughly fact-checking the messages and data provided by the user.

This format is especially useful to cover issues of great social interest, where great diversity of opinion may exist, so as to stimulate greater participation and discussion on the part of the users.

3.2   Short Messages Published by Journalists

It is presented as a news piece composed of brief messages written by the journalist during the development of a topic or event, which can be published in the media communication website or in the journalist's own blog. The origin of the news comes mainly from the investigation and follow-up by the journalist, as well as that of the press agency. In some cases, however, it can also include messages sent by other users in social media like Twitter or Facebook, among others.

Its use is effective in breaking news, since it quickly and concisely reports the known details about the issue. The need for immediacy in updating the data, nevertheless, increases the risk of committing verification and spelling errors. This is its main weakness.

3.3 News Summaries Through WhatsApp or Instagram Stories

Consisting of a summary of the most important news that has occurred throughout the day or week, it depends on the volume of news happening in the city where the media is based. Its approach is based on sending or publishing between three and five brief headlines, along with a link to the full post, which allows the user with scarce time to know the most relevant news. The use of this format in two of the social media with the greatest growth in recent years, WhatsApp and Instagram, allows the media access to a large number of people, especially the young. In addition, the inclusion of the link to the piece of news makes it easy for those users who wish to, to access the complete story in a simple way, thus increasing the number of web site visits.

The main limitation is the condition that WhatsApp users must subscribe to the media mailing list to receive the news.

3.4 News Coverage via WhatsApp or Telegram

This format allows the user to follow an event or news through these instant messaging platforms as a journalistic chronicle. The quick updates this type of coverage requires means it is based on the use of short messages and a few visual resources, intended only to supplement or amplify the information.

The functionality of WhatsApp and Telegram allows messages to arrive directly to the users' mobile devices, so it is important to limit the sending of messages to avoid information overload. The use of this media is only possible after users have signed up to the WhatsApp mailing list or the Telegram broadcasting channel.

Its applicability is broad and it allows local media to be able to offer audiences a useful service for their day to day news or for certain events which they cannot attend, but that are of keen interest to the general public.

3.5 Teaser

This format, originating in the field of advertising, is used to awaken the curiosity and intensify the intrigue of the audience. The most widely used media format consists of short videos that anticipate a few details that keep you guessing what the subject matter or the main protagonist of the piece of news may be. It is usually spread days before the publication of the news or feature to generate buzz.

Its brevity and audio-visual nature make it a format completely adapted to social networks, where the use of eye-catching visual resources is widely accepted on the part of the users. Similarly, its diffusion in this space allows, from the comments and other mechanisms of interaction it receives, the communications media to be able to foresee if the piece will create interest or generate debate in the audience. It is usually reserved for particularly relevant or innovative news or reports.

3.6 Videos Using Instagram IGTV

IGTV is a tool similar to a television channel built into the Instagram application with which users can publish and consume audio-visual content. Unlike another known resource from this social media, the stories, these videos can last up to ten minutes (60 min if the profile is verified by Instagram) and remain stored in the user profiles that publish them, so they can be consulted or seen again over time.

Its vertical format and high sound and image quality are useful to the local media in broadcasting their reports, interviews, featurettes, tutorials or other news videos through mobile devices.

Moreover, IGTV allows users to leave comments that enhance both the interaction between the medium and its audience, as well as among the followers themselves, thus creating a community that shares the same tastes and interests. This enhanced public loyalty is key to local media.

4 Interactivity

Interactivity has a double meaning. On the one hand, it consists of creating several reading routes or itineraries. In this formula, the information consumption is no longer necessarily linear. The public has the freedom to choose how it wants to consume news content. On the other hand, the interactivity makes reference to the public participation through comments, "likes" or shares. These formulas allow the user to express their opinion, but also to create a much more direct relationship with the local media to articulate communities of users.

4.1 Interactive Reports

Inspired by the philosophy of 'choose your own adventure' books, this format allows the public to be able to freely choose the information it wants to consume and in the order it wants to do so. Some of these stories arise in the form of games with a series of questions that guide readers through different scenarios, based on gamification.

The diverse reading routes are achieved by combining links and multimedia resources such as videos, animations or infographics. The format also requires working with different multimedia resources such as audio, images or text. It can also include content coming from social media or invite the public to use the platform for story-related purposes.

Its features make it particularly useful in investigative journalism or data reporting, as it allows information to be presented in a simple, visual and attractive way, thus facilitating audience utilization.

4.2 Chat Meetings

This format consists of digital encounters between an interviewee, users of a chat or social media and a journalist. The latter acts as an intermediary between the interviewee and audience. It is characteristically highly digital because, although the interviewee and the journalist are in the same physical space, the interview is through different digital communications channels such as the media website, a forum, etc. They are also collaborative interviews since, although the journalist proposes questions to talk about, the questions and comments that users send previously or live are its main ingredient.

This formula energizes the community of followers and increases their engagement, since they abandon their passive role in the consumption of news and participate actively by proposing the themes and issues to be dealt with throughout the interview. In addition, it is an attractive format for the interviewees, since it allows them closer and more direct contact with users, but under the moderation of a journalist.

4.3 Voice Notes Using WhatsApp

Considered one of the most widely used features of WhatsApp, this format makes it possible to send short or long duration audios, easily and immediately. Its use allows the news to be complemented with statements by witnesses or other relevant sources to an event, so lending the news greater credibility. They are also relevant to giving users a voice, who can share opinions on a theme proposed by the medium.

This format is particularly useful in local journalism for approaching sources that are difficult to access, since in many cases there are insufficient economic or material resources to travel to the location. The use of this format, however, requires the exhaustive task of identifying the sources and controlling its content to prevent fake news.

4.4 Surveys, Questions and Questionnaires Through Instagram

Surveys, questions and questionnaires are three formats available in Instagram Stories and they serve to encourage user participation. Their use is particularly interesting for the local media, as it allows them to get the public's opinion on various current issues. In addition, knowing what the views of the followers are makes it possible to identify who the target audience is and what their interests and concerns are. This is something that can be combined with big data techniques to create user profiles.

The functionality of these formats means that users can interact and feel involved in the news production process. In this sense, the media can use these formats to ask their followers what topics they would like to have more news on. Moreover, through questions, surveys or creative questionnaires, the media can create sets of questions that inform while entertaining users, activating loyalty strategies.

5 Multimedia

This category brings together those formats that combine multiple forms of expression to present or disseminate news. These major resources mainly include text, image, sound, video, or animation.

5.1 Media Reports

A media report is an informative piece that incorporates all the features of the digital environment such as hypertextuality (non-sequential linking of text, image and sound), interactivity (the ability to relate to users) and multimedia technology (the combination of several resources such as, for example, image, text and sound).

Its objective is to offer an in-depth account of the news, including large numbers of detail, but in an attractive and visual way. By its nature, its format is very versatile and applicable to any subject type. Its use is particularly interesting in questions that involve a lot of data, many protagonists or that require detailed coverage. Its main limitation is cost, which can be high for local media with limited budgets.

5.2 The Expander

This format serves to broaden the news on a particular concept, fact or element that is within the journalistic story. It can be a complement to the multimedia reports. The media offers its readers a kind of link through which they can learn more details about a question that requires in-depth explanation, but without leaving the home page.

The objective of this format is to make it easier for users interested in expanding their knowledge on the subject to be able to do so. So, they might read only the story or better understand its most relevant details by using the expander. It is a useful element in reports on complex issues, which include a lot of data, whose full explanation in the content of the report could hinder the public's understanding of the information.

5.3 Atmosphere

This is a format which accompanies the news with sounds whose goal is to create an appropriate atmosphere for reading. It is, therefore, a multimedia complement, as it extends and enhances the user experience, who may feel part of the reality reported in the news. It can feature music, ambient sounds, a recording or a statement made in the same place as the events. The choice of this format, however, must be done carefully since a very shocking or out of context audio may be a distraction for the users and provoke a loss of interest in regard to the news content.

5.4 Facebook Instant Articles

This format, integrated into Facebook, allows journalistic content to be consumed from the same social media. Its employment by users is simple, since they can quickly access the full articles in a highly visual interface, as well as share news they consider interesting with their friends. A factor that can lead to an increase in the notoriety of the contents, since sharing articles through this platform makes them more viral and, thus, increases their spread of circulation.

The format, moreover, opens the contents of the local media to a potential audience composed of all active Facebook users. Likewise, it enables them to build loyalty to the page of the media in this social network from subscribers interested in easily accessing the contents. The main problem of this type of format, however, is that it's possible that users do not access the media webpage, limiting themselves to only consuming the news that this shares through this application.

5.5 Podcasts

This format consists of the publication of a periodical digital audio or video and can be consumed asynchronously through a webpage, blog or social network, or downloaded to be heard or viewed on computers or mobile devices. The audience may consume the content at any time or place. Podcasts usually offer specialized content intended for an audience interested in the subject. For this reason, their inclusion in social media leads to greater interaction on the part of the users, who can give their feedback or raise new topics or issues to journalists.

Its creation is simple and economical, since it does not need any specific recording and editing software nor specific expertise. Furthermore, there are currently a multitude of technological tools and platforms for storing and sharing this content.

6 Image

The great growth in numbers of users of social media such as Instagram or YouTube has placed audio-visual content in a preferential position. Currently, the use of photographs, videos, graphics and other elements of a visual nature are a claim for the internet audience and are increasingly demanded by users. Being able to view the elements that make up the news allows the public to immerse themselves within the events journalists report.

6.1 Infographics

The infographic is an informative visual representation on a topic. Its aim is to simplify complex ideas and facilitate the understanding of the most significant informative aspects. It is especially useful in very specialized subjects where technical language is employed, or in matters in which a multitude of data or protagonists is involved.

Its approach can be static if it consists only of text and illustrations, icons or graphics; or interactive, if it includes videos, audios, animations, or links through which to consult greater information. This latter modality encourages greater interaction on the part of the reader, who chooses the data they want to expand as they progress with the visualization and understanding of the news provided by the infographic.

This format is directly related to big data since, thanks to the information available in social media, often published by users, interactive maps can be created. Also, thanks to the activity data in social media, maps can be prepared where the news scenarios appear geo-located. Infographics and visualization can make sense of large-scale datasets and provide journalistic meaning.

6.2   360° Image and Video

360° images and videos are two formats of immersive journalism. Through them, the public can focus on a scene from all directions and possible angles, moving freely through the scenario on show. Thus, one of the main attractions that these formats offer is that the audience can decide which part of the reality they want to see and experience the events or situations that occur in it in the first person. Currently, social networks such as Facebook already allow the taking and publishing of this type of image from within the application itself. Likewise, phones, tablets and computers are already adapted to the quality demanded in these formats and allow them to be uploaded and viewed normally.

However, its use at the professional level is still a challenge for the local media, especially for those owning fewer resources. Using these formats requires sizeable economic investment for both the recording devices as well as the editing programs. Additionally, they need specialized professionals with knowledge in virtual reality and 3D animation. This can create an obstacle for some local media when it comes to applying these formats.

7   Streaming

Streaming encompasses the formats that narrate events minute by minute or connect live with the scene of events. These formats, originating in radio and television, have recently been incorporated as a regular resource in the digital press, often through social networks, which have specific tools for broadcasting live.

7.1   Live Broadcasts Through Social Media

Live streaming through social media enables a quick and easy connection to any point where the news is found. This format allows users to follow events at the same time as they are happening, but with the particularity of actively being able to take part in its development, by simultaneously publishing comments. With this formula the media may interact with its followers. Thus, community participation may be encouraged, raising questions that can be answered live or finding out impressions and what is generating most interest between users at the same time as events take place.

Currently, social media such as Facebook, Instagram or YouTube feature free built-in tools to make live video retransmissions. This has potential for local media, which does not have to face any additional economic costs for its use. In this way, any journalist with a mobile device with an internet connection can retransmit any event. Even, those of a local or hyperlocal nature, which are often left out of the news coverage due to a lack of sufficient human or material resources.

7.2 Liveblogging

This format combines the essence of streaming with the practice of blogging. This allows the construction of the journalistic story as a blog post. The greatest potential of live blogging is the real-time transmission of the comments and the analysis of current events in chronological order, in combination with such resources as, for example, summaries, quotes, links, social media posts or multimedia elements.

In this way, users can continue to follow an event live that they have not been able to personally attend. At the same time, they can converse with other users. It also counts on a space for debate and discussion. In this context, the possibility arises for the media to boost its commitment from its followers, inviting them to contribute their point of view on the event and involving them in the process of news production.

Some applications, such as ScribbleLive, make it possible to make extensive journalistic coverage, including images, videos, links, user-aimed surveys and spaces to leave comments. Once the event is over, it gives the option to keep the full coverage, which makes it easier for the public to see the contents afterwards.

8 Transmedia

This category brings together those formats that transcend the space of a single platform and expand the news through various broadcasting channels (newspapers, television, radio, web, social media, etc.), combining different languages and formats (text, image, video, audio, infographics) and encouraging the participation of the public. The aim of these formats is to build several stories about the same event, that are interrelated, but which may be consumed on their own without that affecting the users' comprehension of the content.

8.1 Transmedia Reporting

This format is based on the construction of journalistic content across different platforms. This type of narration combines different interrelated stories, but at the same time they maintain their independence and make separate complete sense, thus adapting themselves to different styles and perspectives according to the channel selected for their diffusion.

Although the digital context, with its many social media, websites and blogs, has become the preeminent space of this type of format, it can overcome virtual barriers and extend to the real world. Thus, part of the narratives it relates can be translated into offline actions like books, conferences, educational projects, photographic exhibitions, contests, etc. The fact of combining different perspectives, according to the languages, formats and channels used, makes one of the main virtues of this format: the capacity to enrich the news story, making it more immersive, integrating and participatory.

It is a format that offers a multitude of possibilities to local media, who using the online and offline space can expand the reach of their news coverage, establishing greater contact with their audience and inviting it to actively participate in its contents. It is, however, a complex and expensive format that implies knowing the language of each platform (television, radio, digital media, blog, social media) and expressive resource (video, podcast, text, hypertext, streaming), to adapt the different stories that make up the universe surrounding the news. Besides, it is a format that needs a broad theme or event with sizeable, amounts of data and stories to tell, which implies dedicated research and production time. This is an obstacle for local media with scarce economic resources.

8.2 Collaborative Reporting

This format is based on reporting produced between journalists and citizens. Its approach is based on the proposal of a topic by journalists, who through social media ask their followers to send them information, statements or ideas related to the subject. Likewise, they can ask the users to be the ones who suggest topics or ideas they find relevant or controversial within their city, that usually do not feature in the national media.

Unlike other formats, citizen participation is not limited to commenting or disseminating journalistic content through different platforms. On the contrary, citizens participate actively and directly in the creation of content. In this way, the media can count on multiple sources of information that allow the enrichment of content and the raising of awareness of issues that would otherwise be left out of the public agenda.

For this reason, its success depends directly on the initiative of the users. If they do not send their proposals, comments or content the format fails for lack of collaboration.

9 Conclusions: New Formats and Big Data to Ensure the Future of Local Journalism

Local journalism faces an uncertain future (Nielsen 2015) due to the impact of digital technologies that are structurally transforming the sector. Faced with this situation, it must place its bets on innovation (Küng 2015) to reconnect with the public and offer attractive content of a civic character that adds value (Casero-Ripollés 2014). These may be able to redefine the business models, that were strongly weakened by the converging economic crisis and digitization (Picard 2014).

With the aim of producing news adapted to the digital environment that is better able to connect with the audience, six broad categories of formats based on social media for local journalism have been identified. These are: storytelling, interactivity, multimedia, image, streaming, and transmedia. In turn, these six categories are divided into twenty-one formats among which are, including others, live blogging, the tweet magazine, summary information through WhatsApp or Instagram Stories, Facebook Instant Articles, the expander, the teaser, collaborative reporting, surveys, questions and questionnaires through Instagram or 360° image and the 360° video.

Many of these formats incorporate the use of big data for local journalism, improving its potential, when it comes to collecting, processing, analyzing, compiling and disseminating information. These formats make up a catalogue that represents a wide and varied range of options that can allow local journalism to introduce innovation in news production and, with it, to regain the confidence of the public, ensuring its survival and future in the digital scenario.

Acknowledgements This work is part of the research project UJI-B2017-55, funded by the Universitat Jaume I of Castelló, within the Plan of Promotion of Research (2017).

References

Campos-Freire F, Rúas-Araújo J, López-García X, Martínez-Fernández VA (2016) Impacto de las redes sociales en el periodismo. El Profesional de la Información 25(3):449–457. https://doi.org/10.3145/epi.2016.may.15

Casero-Ripollés A (2010) Prensa en internet: nuevos modelos de negocio en el escenario de la convergencia. El Profesional de la Información 19(6):595–601

Casero-Ripollés A (2014) La pérdida de valor de la información periodística: causas y consecuencias. Anuario ThinkEPI 8:256–259

Firmstone J (2016) Mapping changes in local news. Journalism Pract 10(7):928–938. https://doi.org/10.1080/17512786.2016.1165136

García-Avilés JA, Carvajal-Prieto M, Arias-Robles F (2018) Implementation of innovation in Spanish digital media: analysis of journalists' perceptions. Revista Latina de Comunicación Soc 73:369–384

Hess K, Waller L (2016) Local journalism in a digital world. Palgrave, London

Jarvis J (2014) Geeks bearing gifts: Imagining new futures for news. CUNY Journalism Press, New York

Kalogeropoulos A, Nielsen RK (2018) Investing in online video news: a cross-national analysis of news organizations' enterprising approach to digital media. Journalism Stud 19(15):2207–2224. https://doi.org/10.1080/1461670X.2017.1331709

Kim YC, Ball-Rokeach SJ (2006) Community storytelling network, neighborhood context, and civic engagement: a multilevel approach. Hum Commun Res 32(4):411–439

Küng L (2015) Innovators in digital news. IB Tauris, London

Lewis SC, Westlund O (2015) Big data and journalism: epistemology, expertise, economics, and ethics. Digit Journalism 3(3):447–466. https://doi.org/10.1080/21670811.2014.976418

López-García X (2008) Ciberperiodismo en la proximidad. Comunicacion Social Ediciones y Publicaciones, Sevilla

Negreira-Rey MC, López-García X, Rodríguez-Vázquez AI (2018) Los cibermedios locales e hiperlocales en España y Portugal. La fase de búsqueda de modelos. Sur le journalisme 7(2):50–63

Newman N (2009) The rise of social media and its impact on mainstream journalism. Reuters Institute for the Study of Journalism, Oxford

Nielsen RK (2015) Local journalism: the decline of newspapers and the rise of digital media. IB Tauris, London

Picard RG (2014) Twilight or new dawn of journalism? Evidence from the changing news ecosystem. Journalism Stud 15(5):500–510. https://doi.org/10.1080/1461670X.2014.895530

Salaverría R (2019) Digital journalism: 25 years of research. Review article. El Profesional de la Información 28(1):e280101. https://doi.org/10.3145/epi.2019.ene.01

Vázquez-Herrero J, Negreira-Rey MC, López-García X (2018) Transmedia discourse at the local and hyperlocal sphere. In: 13th Iberian conference on information systems and technologies (CISTI). IEEE, pp 1–5

Andreu Casero-Ripollés Professor of Journalism and Dean of the School of Humanities and Social Sciences at Universitat Jaume I (UJI). Previously, he was head of the Department of Communication Sciences and director of Journalism Studies. He holds a degree from the Universitat Autònoma de Barcelona and a Ph.D. from the Universitat Pompeu Fabra. He is a member of the Institut d'Estudis Catalans. He works on the transformations of digital journalism and political communication.

Silvia Marcos-García Ph.D. in Communication Science, a degree in Journalism and a Master's in New Trends and Processes of Innovation in Communication from Universitat Jaume I (UJI). Her lines of research focus on political communication and journalism on social media. Specifically, her work is based on the study of the use of social media such us Twitter, Facebook and Instagram by political actors, media and citizens.

Laura Alonso-Muñoz Ph.D. in Communication Science and Postdoctoral Research fellow from the Ministry of Economy, Industry and Competitiveness of Spain in the Department of Communication Sciences at Universitat Jaume I (UJI). She has a degree in Journalism and a Master's in New Trends and Processes of Innovation in Communication from UJI. She also is graduated in Political Science and Administration from the Universitat Pompeu Fabra. She works on the transformations of digital journalism and political communication.

Information Visualization and Usability: Tools for Human Comprehension

Ángel Vizoso

Abstract This chapter makes a review on the progress of two growing areas in the field of communication: information visualization and usability. Information visualization has experimented a lot of changes during the last few years, especially with its arrival to the Internet and the development of different narrative forms by taking advantage of the main characteristics of the new environment. Something similar has occurred with the concept of usability. The following sections explore the significance of this young idea whose spread took place especially with the start of the Internet. With the objective of completing this theoretical framework, a small usability test was conducted with five renowned visualizations. The results of this test show that even the most salient journals do not fulfill many of the main usability recommendations. Hence, this area still has a considerable way to go regarding its appliance to information visualization.

Keywords Information visualization · Usability · Infographics · User-centered design

1 Milestones and Development of Information Visualization

Information visualization has been one of the areas with a major development in the field of communication. Its main goal is the presentation of information and data in a comprehensive and orderly way (Uyan Dur 2014). Thus, this discipline can be defined as "the representation and presentation of both data and information that takes advantage of our visual capability in order to expand our knowledge" (Alcalde 2015: 20). Likewise, it is "the use of visual representations to explore, make sense of, and communicate data" (Few 2014: 2). As noted through these definitions, there have been many efforts to define and delimit the area, and most of them highlight the power of information visualization as a communicative tool for the knowledge enrichment (Olmeda-Gómez 2014).

Á. Vizoso (✉)
Universidade de Santiago de Compostela, Santiago de Compostela, Spain
e-mail: angel.vizoso@usc.es

© Springer Nature Switzerland AG 2020
J. Vázquez-Herrero et al. (eds.), *Journalistic Metamorphosis*,
Studies in Big Data 70, https://doi.org/10.1007/978-3-030-36315-4_7

When talking about information visualization we are analysing a booming area, especially in journalism. Thus, it is not a new area or an exclusive communicative tool of our days (Figueiras 2014). It is possible to find antecedents of this visual genre in books like William Playfair's *The Commercial and Political Atlas* and *Statistical Breviary* published in 1786 and 1801 respectively. As pointed out by Cairo (2008), these books contain the first examples of bar, line and pie charts. A few years later, in 1854, it is possible to find another great milestone in the development of information visualization. The British doctor John Snow showed the geographical spread of more than five hundred deaths due to cholera during ten days in the Soho district, London. Snow's map combines two datasets. First of all, the number and location of the cholera deaths and, secondly, the water sources location. Hence, thanks to this map it is possible to find out the relation among the deceases and the contamination of some water sources in the neighbourhood. Edward Tufte highlighted the importance John Snow's contribution by opening a new way of providing "direct and powerful testimony about a possible cause-effect relationship" (1997: 6).

Less than a decade later, the French engineer Charles Joseph Minard published one of the most iconic visualizations ever, a chart of the losses of Napoleon's army during its Russian Campaign (1812–1813). Minard's chart shows the development of different variables: "the size of the army, its location on a two-dimensional surface, the direction of the army's movement and temperature in various dates during the retreat from Moscow" (Tufte 2001: 40). All of that in a very visual chart which reflects the regression of Napoleon's army both during its advance towards Moscow first and its withdrawal then.

Although the examples previously noted are considered a key part of the development of information visualization, they were not published by the press, at least, initially. It exists a deep debate among some scholars about the start of press infographics. Thus, many of them consider that the description of Admiral Vernon's attack to Portobello's bay is the first example of infographics available in the printed press (Sullivan 1987 in Franco 2005). It was published by the *Daily Post* on the 29th of March 1740. However, it exists broad agreement among scholars in identifying the graphic published by the British diary *The Times* on the 7th of April 1806 about Isaac Blight's homicide as the first example of infographics in press (Peltzer 1991). This graphic contains a draw of the view of Mr. Blight's house as well as a plan of its inside with an explanation of the itinerary followed by its murders (Cairo 2008).

However, information visualization experts tend to coincide in highlighting the launching of the newspaper *USA Today* as one of the main milestones in the development of this field. Launched in September 1982, *USA Today* changed the way in which press journalism was produced by introducing a new design idea based on the use of colour instead black and white, an increase in the use of images and a clear commitment in the use of infographics (Lallana 1999). After the emergence of this journal, some global events like, for instance, the Gulf War and the difficulties for accessing audiovisual materials during its course, had a great influence in the development of this genre. The media tried to soft the shortage of videos and pictures of this conflict by using visualizations to explain both the development of the war as well as the weapons and materials used by the armies.

1.1 From Printed to Interactive Infographics

As pointed out before, information visualization has its origin in many works related with disciplines like economics or demographics. Then, the genre reached a new characterization with its adoption by the media and, more concretely, by the printed press. This integration resulted in the development of printed infographics, a particular way of informative communication which can be defined like "an informative contribution based on both iconic and typographic elements, which allows the comprehension of facts, actions or current topics or, at least, some of their main aspects by accompanying or replacing an informative text" (Valero-Sancho 2001: 21).

Following Cairo (2008), infographics are not an ornamental object with the objective of make the pages of the newspaper more light, dynamic or colourful. It has to work as a tool for the reality analysis improving readers' comprehension.

Through these two definitions it is possible to assess two ideas. Firstly, the nature of infographics as a journalistic genre, similarly to others like reports or interviews. Secondly, the power of infographics as a vehicle for information transmission. Moreover, concerning its production, it has experimented an evolution during the last two decades. Nowadays infographics are increasingly the result of the work of journalists with design and data management abilities while, in the past, it was produced by artists who adapted its know-how to the needs of journalism (Cairo 2017).

The advance of infographics since the 90s has been continuous, as noted before. All of that thanks to the inclusion of new visual forms and technological tools which have allowed the adaptation and improvement of this genre. In addition, these changes have led into the birth of a new form of visualization, multimedia infographics, considered a completely new communicative form by many scholars (Arévalo 2009).

Wibke Weber, defines an interactive information graphic as "a visual representation of information that integrates different nodes into a coherent whole and offers at least one navigation option to control the graphic" (2013). The youth of the use of this genre by the online media has provoked the emergence of a wide range of ways to name it. Thus, leveraging its characteristics—multimedia, hypertext, and interactivity, a lot of nomenclatures have emerged. Some of them are, for instance: digital infographics (Pinto Rodrigues 2012), online infographics (Nogueira 2018), interactive infographics (Dick 2013), or multimedia infographics (Salaverría and Cores 2005). Another ones are data visualization (Iliinsky 2012) or information visualization (Anderson 2017). However, it exists a deep debate among scholars. While some of them consider that infographics are a subdiscipline of information visualization, others think that both infographics and information visualization are part of the same reality.

Despite these diverse ways to name it, which is undeniable is that multimedia infographics have experimented a huge advance in the last few years. This advance has its explanation in the fact that, while printed infographics do not rely on a close connection with technology, multimedia does. Thus, it is possible to notice different stages in the development of multimedia infographics which, following Gomes-Amaral (2010), are:

- *Content replication.* It was the initial stage of the publication of infographics online. Therein, newspapers replicated the same content published before on their printed version without any interactivity or multimedia capability.
- *Addition of certain interactivity.* The second stage was marked by the integration of a certain level of interactivity in the infographic pieces previously published in the printed edition of the media.
- *Inclusion and combination of multimedia elements.* The main characteristic of this third stage is the use of audio, video, text, images and hyperlinks in a product conceived for its publication online.
- *Data based infographics.* Visualizations linked to a database with real-time updating. This fourth stage allows the highest level of personalization and exploration for the reason that each individual use can refer to a specific part of the data in order to obtain an individual and exclusive experience.

Hence, when talking about infographics—both printed and multimedia—we are talking about a flexible and efficient journalistic genre and communicative tool which can be a complement for the information or present it autonomously (Zwinger and Zeiller 2016). In addition, infographics have demonstrated its capability to deal with data from many disciplines, and it is the reason why its use is becoming more frequent in disciplines like art (Li 2018) or economy (Gatto 2015).

Although the development of information visualization was steady during the last twenty years, further challenges have emerged during the last decade. The onset of new devices like mobile phones or tablets provoked new needs in the production of news graphics, especially the journalistic ones. Journalists, designers and programmers who carry out the task of data extraction and content production for these news products have now the need of adapting its visual appearance and its functions to these different ways of consumption. Hence, concepts like responsive design or usability play now a central role when producing information visualization.

2   The Concept of Usability and Its Relevance for Information Visualization

Usability has been defined by the International Organization for Standardization (ISO) as "the extent to which a product can be used by specified users to achieve specified goals with effectiveness, efficiency, and satisfaction in a specified context of use" (International Organization for Standardization 2018: 15). Although this agency opens its definition to all products, the historical development of the concept has been centered mainly in computing tools in both software—computer programs, video games or, more recently, mobile apps, and hardware sides—computers in its different variants, mobile phones or tablets among other devices.

In 1993 Jakob Nielsen talked about the growing importance of user interfaces in the computing of that time. Nielsen pointed out that the changes in the use of computers—from specialized users to the popularization of these machines—lead to the need of "making life easier for the user" (Nielsen 1993: 8). Hence, in the last years the concept of usability is more related with realities like the "learning agility and the ease of use of a product" (Rache et al. 2014: 180) and linked very closely to the idea of the user-centered design (UCD) where one of the main focus when producing a technological product is the capability to adapt it to the final user (Mayhew 1999). Regarding this idea, Nielsen (1993: 26) established five attributes to usability:

- *Learnability*. Systems should be easy to learn in order to make possible a quick adaptation of the user.
- *Efficiency*. This attribute refers to the ease of learning how the product works in order to increase users' knowledge.
- *Memorability*. This characteristic points out the need of making systems whose use would be easy to remember. Thus, when users come back after a certain period of time without using it, it would not be necessary to learn how it works.
- *Low error rate*. Errors should not appear or, at least, their appearance should be scarce. Therefore, the system should be capable of recovering itself from an error.
- *Satisfaction*. Nielsen pointed out that systems should be pleasant to use, a feature that would lead the users to a major satisfaction.

As it is foreseeable, the major development of both the theory about this concept and the design of interfaces and devices with the most recommended usability criteria has been taking place from the beginning of the Internet, democratisation of computers and, in the last few years, mobile phones (Souza and Maciel 2018).

2.1 The Challenges of Usability Evaluation

In the communicative context of our days, "users are surrounded by a broad range of networked interaction devices" (Yigitbas et al. 2018: 231). This fact has provoked a lot of changes in the production of communicative products and tools. However, one of these changes stands out among the others: the user has become a central part in the designing process. This phenomenon is what experts call the user-centered design, "an iterative design process in which designers focus on the users and their needs in each phase of the design process" (Interaction Design Foundation 2019). Hence, at the present time, those professionals who produce tools and products both in the area of communication or in other ones have to put the final users in the middle of their projects in order to provide them with efficient and attractive experiences.

In the course of this designing process as well as after its launching, both these teams and some outside bodies conduct different usability tests and experiments. These evaluations are used to assess the efficacy and the pertinence of both the products and the techniques employed in its production and can be performed in different ways.

For instance, Rache et al. (2014) explained the existence of two types of methods for the evaluation of usability. First of all, there would be techniques where the presence and participation of the users are needed. These could be:

- *User testing.* In this method, the users are required to interact with the interface in both free and directed navigation.
- *Card sorting.* Researchers present different cards to the users. They are required to group them in different categories relevant for them for, finally, naming these categories.
- *Interviews.* These could be directed, semi-structured or non-directional. All these three types have the objective of knowing user's perceptions on the use of a product.
- *Questionnaire.* It exists some pre-determined and standardized questionnaires like the Website Analytics Measurement Inventory (WAMMI) or the Usability Measurement Inventory Software (SUMI).
- *Creativity methods.* Example of these techniques could be brainstorming or association of ideas, among others.
- *Critical incident.* Qualitative interview where the interviewed user identifies different events. The person highlights these events, how they evolved and their impact and consequences.
- *Observation.* In this method, researchers are observers of the normal performing of users' activities.

These authors noticed that another possibility is the use of techniques where the presence of the user is not required. These methods are performed only by researchers or experts which can notice the usability particularities of any product by using different procedures:

- *Heuristic analysis.* Inspection of the interface with the objective of detecting the positive and negative aspects of its usability.
- *Cognitive walk-through.* Simulation of the user's cognitive behaviour. The authors identified three phases in the implementation of this method: (1) definition of the scenario and the aims of the study; (2) evaluation phase with questions for each task performed; (3) identification of any possible usability problem by using the answers given to the previous questions.
- *Personas.* Analysis of both the needs and profiles of the potential users of a product with the objective of creating fictional characters. Then, designers can refer to these profiles when producing a new interface.
- *Automated evaluation.* Use of algorithms for the automatic analysis of the quality of the presentation.
- *Evaluation by expertise.* Identification of any usability problem by an expert or a group of experts.
- *Analysis of documents and reports.* Review of documentation from diverse sources. With the use of this method, researchers can make their own judgements on different products.
- *Creativity methods.* As previously noted, this method involves the utilization of brainstorming or the association of ideas and it is valid for research in both the presence and the absence of the users.

3 Method

To complete the theoretical approach presented in this chapter with an empirical experience, a small heuristic usability test was conducted in order to see if some of the most recognized visualizations published in the last few years fulfill the main usability criteria set by different authors. To reach this objective, an analysis card used by de Oliveira and Guimarães (2017) adapted by selecting twenty of the items contained on it. It is necessary to point out that these authors reviewed previous research in order to set up this tool. It is necessary to note that there are different methods for usability evaluation as pointed out by Rache et al. (2014). Then, five visualizations were chosen. These multimedia infographic examples were awarded in the Malofiej Awards' Best of Show—Online from 2015 to 2019. After its identification, the analysis card was applied to these items in an heuristic analysis in order to compare the presence of the chosen criteria in these renowned works. The twenty elements reviewed in this analysis card are listed in Table 1 and were scored from 0 to 5 points. Thus, each visualization could obtain an overall score from 0 to 100 in fulfilling the established items.

The authors are aware that this experience is not representative about the development of the usability criteria or the user-centered design implementation. However, it will be very useful for pointing out some of the key strengths and weaknesses of five significant visualizations awarded by an international jury.

3.1 Selected Works

In this section we will describe briefly the works which integrate our sample, five prize-winner visualizations at the Malofiej Awards, the most relevant recognition for infographics. Malofiej Awards take place at the University of Navarra every year since 1993, thanks to the joint effort of both the University of Navarra and the Spanish Chapter of the Society for News Design, one of the most important associations in the professional field of information visualization. For a few days, professionals and scholars from around the world share their experience in workshops and conferences. However, one of the main events is the awards ceremony where the most salient visualization examples in different categories are recognized. For instance, in 2019 there were prizes for both digital and printed visualizations in categories like 'Best of Show', 'Best Map', 'Climate Change and Environmental Commitment', 'Human Rights', 'Equality and Women's Promotion' as well as the classic 'Gold', 'Silver' and 'Bronze' qualifications for breaking news and features printed and online graphics.

As previously stated, the five online visualizations awarded as the 'Best of Show' from 2015 to 2019 were selected. They will be briefly described before displaying the main findings of our small heuristic usability test.

Table 1 Usability criteria reviewed in the analysis card

Name	Questions about each topic
User control and freedom	Does the user have the freedom to control the visualization?
History	Does the visualization tell a story by itself?
User control	Is the user allowed to control the data?
Orientation and help	Does the visualization provide guidelines about its utilization?
Provide multiple levels of detail	Does the visualization offer the possibility to access multiple levels of detail for the information?
Details on demand	Does the information provide the possibility to ask for any further data?
Spatial organization	How good is the placement of the elements in the visualization?
Grouping and distinguishing items by format	Does the visualization distinguish items by its format?
Grouping and distinguishing items by location	Does the visualization distinguish items by its location?
Help and documentation	Does the visualization provide any documentation or context details about the topic?
Legibility	What is the ease to read the data in the visualization?
Aesthetic and minimalist design	Is the design minimalist?
Information density	What is the density of the provided data?
Cognitive complexity	Is this visualization difficult to understand to a medium user?
Multivariate explanation	Does the visualization use multiple variables for explaining the information?
Put the most data in the least space	Is the visualization efficient in using the space?
Formulate cause and effect	Does the visualization provide the cause and effect of the data or information showed?
Consider people with colour blindness	Does the visualization avoid colours like red and green?
Conciseness	Is the visualization concise by avoiding data redundancies?
Integrate text wherever relevant	Does the infographic use text only when necessary for extend the information?

Source de Oliveira and Guimarães (2017)

Areas Under Isis Control.[1] This visual story, whose last update was on the 28th December of 2015, describes the advance of both the Iraqi Army and the ISIS militants in their fight for the lands of Iraq, Syria and Jordan. The user only has to scroll down in order to find a great variety of maps—up to thirty, charts and pictures of the conflict zone. Although the content does not show any interactivity, *The New York Times* visualization team tried to tell the whole story with the use of charts, maps, text and images.

Unaffordable Country[2] allows the user two possibilities. First, it is possible to reach historical information from 1995 to 2014 about the house price in every single region of the United Kingdom. It is possible to find out the highest, lowest and median price of a house in the chosen area. Then, the users have the option of entering their salary in order to check how affordable could be buying a house with their current earns. All of that in an interactive choropleth map where a chromatic scale from blue to red shows the price of the housing.

Olympic Races Social Series.[3] The Olympic Races Social Series were a set of visualizations published by *The New York Times* during the 2016 Rio de Janeiro's Olympic Games in order to give an account of the development of the athletics or swimming races among others. There were explanations about the most successful countries in each discipline. All the visualizations followed very similar patterns. They were built-in news stories as an animated content that the users could play. Then, a simple and lineal representation of the swimming pool or the running track was presented with an iconic draw of the athletes and its distance and position during the course of the races. At the end, the visualization showed which athlete won the gold, silver and bronze medal as well as the distance between the first competitor and the rest of the participants. The main point of these items is that they were shared as GIFs through *The New York Times'* social media accounts, being a very visual content whose simplicity and clarity could attract more readers to the whole information, which was completed with text, images and, sometimes, other graphic examples.
Although the whole series were awarded, the authors selected one example of this set, the news piece about Usain Bolt's victory in the 100 m race published on the 14th of August 2016.

The Science of Hummingbirds.[4] This visualization has the objective of explaining the particularities of hummingbirds, one of the smallest bird species. In this work, the

---

[1]Published by *The New York Times* in 2014 and awarded in 2015. Available at https://www.nytimes.com/interactive/2014/06/12/world/middleeast/the-iraq-isis-conflict-in-maps-photos-and-video.html.

[2]Published by *The Guardian* in 2015 and awarded in 2016. Available at https://www.theguardian.com/society/ng-interactive/2015/sep/02/unaffordable-country-where-can-you-afford-to-buy-a-house.

[3]Published by *The New York Times* in 2016 and awarded in 2017. Available at https://www.nytimes.com/2016/08/15/sports/olympics/usain-bolt-100-meters-justin-gatlin-results.html?smid=pl-share.

[4]Published by *National Geographic* in 2017 and awarded in 2018. Available at https://www.nationalgeographic.com/magazine/2017/07/the-science-of-hummingbirds/.

user can access a lot of information about these birds like their speed, comparisons about their size, the particularities of their tongue or their brain. All of that with a combination of real images, draws, text, and charts in order to provide a lot of information on these creatures.

A Window into Delhi's Deadly Pollution.[5] This visualization is the result of the placement of a camera in the top of a building in Delhi in order to advert the pollution levels in the city. The resulting story was composed by a map with the location of the camera and the presentation of hundreds of images from that point. The user has to scroll down the piece to make a journey across the pollution levels between the 29th of October and the 8th of November 2018. The story offers comparisons between different hours of the day as well as charts with data collected from pollution measure stations in the city.

4   Findings of the Usability Test

As explained in the method section, five renowned visualizations were reviewed in order to find how they apply some of the most common usability characteristics. Thus, in this section the main strengths and weaknesses of each visualization will be highlighted by pointing out those areas where each work is salient as well as noting those ones were not. Therefore, before explaining the singularities of each visualization regarding its usability patterns, the final score of each test will be showed in Table 2. It is necessary to note that due to the smallness of the sample this will serve just for monitoring some trends and especially for making a slight comparison among different visualizations which obtained the same award.

As shown in Table 2, *The Guardian's Unaffordable Country* reached the highest score in our usability test. This visualization obtained the maximum rating of five points in the following categories: user control and freedom, user control, details on demand, spatial organization and formulate cause and effect. Compared with the

Table 2   Final results of the usability test

Title	Publication	Year	Final score
Unaffordable Country	*The Guardian*	2016	83/100
Areas Under ISIS Control	*The New York Times*	2015	77/100
The Science of Hummingbirds	*National Geographic*	2018	75/100
A Window into Delhi's Deadly Pollution	*Reuters*	2019	70/100
Olympic Races Social Series	*The New York Times*	2017	49/100

Own elaboration

---

[5]Published by Reuters in 2018 and awarded in 2019. Available at https://graphics.reuters.com/INDIA-POLLUTION/01008173281/index.html.

other four examples, the user can control which data are displayed and it offers the highest possibilities when looking for the data of a particular area. It is a good example of spatial organization because the map is optimized for an efficient displaying in any screen without any infill material. The lowest scores for this work were the three points obtained in both history and cognitive complexity. Compared with the rest of visualizations reviewed, the story narrated in this one is not as evident as the other ones. Something similar occurs with its cognitive complexity. A certain level of literacy in the use of interactive maps is needed for navigating through this work.

The highest values for *Areas Under ISIS Control* were reached in the following items: history, provide multiple levels of detail, grouping and distinguishing items by location, and formulate cause and effect. Due to the extent of this example and as in contrast with the previous one, the narration of a whole history in this visualization is a good example of how this journalistic genre can be combined with elements like images, videos or text in order to narrate complex data and facts. Thus, in this case, *The New York Times* used a cascade of maps and charts that, together with text and images, completes a whole story which formulates the causes and effects of the described scenario by itself. In sum, these materials are correctly grouped both by its format and its location, which simplifies the ease of its read. However, the lowest values for this example were given in areas like its informative density—it constitutes an in-depth visualization with a lot of information—or the lowest possibilities for its control by the user. Moreover, this visualization does not have any interactive possibility at all.

National Geographic's The Science of Hummingbirds obtained a very similar qualification if compared with the previous example. This is thanks to its salience in criteria like the fact that it narrates a whole story by itself—history—or different qualities related to its design. For instance, this visualization uses text only wherever it is relevant, considers people with colour blindness and puts the most data in the least space. However, it obtained a score of two points in areas like control possibilities for the user or the existence of the possibility of obtaining data on demand.

A Window into Delhi's Deadly Pollution had a score of 70 points. This visualization combines the highest score obtained in fields like the integration of text wherever it is relevant or its conciseness with the fact that users do not have any control possibility, or they do not have the opportunity of obtaining details on demand. It is a good example of what some experts call 'scrollytelling'. Scroll is the main action that users are allowed to perform in this visualization without any other exploration possibilities apart from those ones established by the designers.

Finally, *The New York Times' Olympic Races Social Series* obtained the lowest qualification in our study. This is due to the nature of the visualizations, designed for being played without any possibility of interaction. Therefore, although this example reached the highest rates in areas like considering people with colour blindness or the ease of its comprehension by any user, all those categories related with the level of details provided or the control options available for users had low scores. Then, watching these results, it is possible to notice that this visualization was envisaged with a supplementary and visual function, not as the main way for displaying the information.

5 Conclusions

In sum, the aim of this chapter was to highlight the growing importance of usability in the present context of information visualization. As noted in the previous sections, information visualization has been one of the journalistic genres with a major development since the 1990s decade. Especially with the arrival of the Internet, this way of communication has explored new narrative forms and capabilities.

However, in this context, usability and efficiency criteria play now a central role. As stated, information visualization teams have to pay more attention than ever to this side of the development. Here, journalists, designers and programmers try to make visualizations where efficiency, attractive and informative relevance have to be present in one single product.

To prove how important is following usability patterns in nowadays' journalism, a small usability test was conducted. This brief experiment has allowed us to appreciate some of the weaknesses and strengths of high-renowned visualization works in order to highlight the importance of considering as much usability criteria as possible when producing informative pieces like these.

Acknowledgements This chapter is prepared within the framework of the project *Digital native media in Spain: storytelling formats and mobile strategy* (RTI2018-093346-B-C33), from the Ministry of Science, Innovation and Universities (Government of Spain). The project is co-funded by the European Regional Development Fund (ERDF). This text is also prepared as part of the activities of Novos Medios Research Group (Universidade de Santiago de Compostela, GI-1641), supported by the programme *Consolidation and Structuration of Competitive Research Units* of the Galician Regional Government (ED431B 2017/48). The author is also beneficiary from the Education's University Faculty Training Programme (FPU), financed by the Ministry of Science, Innovation and Universities (Government of Spain).

References

Alcalde I (2015) Visualización de la información: de los datos al conocimiento. UOC, Barcelona
Anderson C (2017) Social survey reportage: context, narrative, and information visualization in early 20th century American journalism. Journalism 18:81–100. https://doi.org/10.1177/1464884916657527
Arévalo G (2009) La infografía interactiva: Un género por desarrollar. Chasqui 107:64–67
Cairo A (2008) Infografía 2.0: visualización interactiva de la información en prensa. Alamut, Madrid
Cairo A (2017) Nerd journalism: how data and digital technology transformed news graphics. PhD dissertation, Universitat Oberta de Catalunya
de Oliveira MR, Guimarães C (2017) Adapting heuristic evaluation to information visualization—a method for defining a heuristic set by heuristic grouping. In: Proceedings of the 12th international joint conference on computer vision, imaging and computer graphics theory and applications. SCITEPRESS—Science and Technology Publications, Porto, pp 225–232
Dick M (2013) Interactive infographics and news values. Digit J 2:490–506. https://doi.org/10.1080/21670811.2013
Few S (2014) Why do we visualize quantitative data? Visual Business Intelligence. Retrieved from http://www.perceptualedge.com/blog/?p=1897

Figueiras A (2014) How to tell stories using visualization. In: 2014 18th international conference on information visualisation, Paris, France, 15–17 July 2014, pp 18–26

Franco G (2005) La infografía periodística. Anroart, Las Palmas de Gran Canaria

Gatto MAC (2015) Making research useful: current challenges and good practices in data visualisation. Reuters Institute. Retrieved from https://ora.ox.ac.uk/objects/uuid:526114c2-8266-4dee-b663-351119249fd5

Gomes-Amaral RC (2010) Infográfico jornalístico de terceira geração: análise do uso da multimidialidade na infografia. PhD dissertation, Universidade Federal de Santa Catarina

Iliinsky N (2012) Why is data visualization so hot? Visually Blog. Retrieved from https://visual.ly/blog/why-is-data-visualization-so-hot/

Interaction Design Foundation (2019) What is user centered design? Retrieved from https://www.interaction-design.org/literature/topics/user-centered-design

International Organization for Standardization (2018) ISO 9241-11:2018, ergonomics of human-system interaction—part 11: usability: definitions and concepts. Retrieved from https://www.iso.org/obp/ui/#iso:std:iso:9241:-11:ed-2:v1:en

Lallana F (1999) Diseño y color infográfico. Rev Lat Comun Soc 13. Retrieved from http://www.revistalatinacs.org/a1999c/150lallana.htm

Li Q (2018) Data visualization as creative art practice. Vis Commun 17:299–312. https://doi.org/10.1177/1470357218768202

Mayhew DJ (1999) The usability engineering lifecycle. In: CHI'99 extended abstracts on human factors in computing systems, Pittsburgh, Pennsylvania, 15–20 May, pp 147–148

Nielsen J (1993) Usability engineering. Academic Press, San Diego

Nogueira AG (2018) Controvérsias sobre a tipificação e a identidade da Infografia Online como Género Jornalístico. Rev Asoc Esp Investig Comun 5:50–65. https://doi.org/10.24137/raeic.5.9.7

Olmeda-Gómez C (2014) Visualización de información. Prof Inf 23:213–220. https://doi.org/10.3145/epi.2014.may.01

Peltzer G (1991) Periodismo Iconográfico. Rialp, Madrid

Pinto Rodrigues SM (2012) Infografia digital: Expresso e Público, a que distância ficam do New York Times? Prisma.com 19:1–24

Rache A, Lespinet-Najib V, Andre JM (2014) Use of usability evaluation methods in France: the reality in professional practices. In: 2014 3rd international conference on user science and engineering (i-USEr). IEEE, Shah Alam, pp 180–185

Salaverría R, Cores R (2005) Géneros periodísticos en los cibermedios hispanos. In: Salaverría R (ed) El impacto de internet en los medios de comunicación en España. Comunicación Social, Sevilla, pp 145–185

Souza M, Maciel F (2018) Adding eye tracking data collection to smartphone usability evaluation: a comparison between eye tracking processes and traditional techniques. In: Ahram TZ, Falcão C (eds) Advances in usability, user experience and assistive technology. Springer, Cham, pp 272–282

Tufte ER (1997) Visual and statistical thinking: displays of evidence for making decision. Graphics Press, Chesire

Tufte ER (2001) The visual display of quantitative information. Graphics Press, Cheshire

Uyan Dur BI (2014) Data visualization and infographics in visual communication design education at the age of information. J Arts Humanit 3:39–50. https://doi.org/10.18533/journal.v3i5.460

Valero-Sancho JL (2001) La infografía. Técnica, análisis y usos periodísticos. Universitat Autònoma de Barcelona—Servei de Publicacions, Barcelona

Weber W (2013) What is an interactive information graphic? In: Malofiej infographic world summit—infographic awards. Malofiej Graph

Yigitbas E et al (2018) Usability evaluation of model-driven cross-device web user interfaces. In: Bogdan C et al (eds) Human-centered software engineering. Springer, Cham, pp 231–247

Zwinger S, Zeiller M (2016) Interactive infographics in German online newspapers. In: Aigner W et al (eds) Proceedings of the 9th forum media technology, FMT 2016 and 2nd all around audio symposium 2016, St. Pölten, Austria, 15 Nov, p 54

Ángel Vizoso Ph.D. Student at Universidade de Santiago de Compostela's Contemporary Communication and Information Ph.D. programme and member of Novos Medios research group at the same university. His research is focused mainly in the area of information visualization, fact-checking and journalistic production for online media. He is also beneficiary from the Education's University Faculty Training Programme (FPU), financed by the Ministry of Science, Innovation and Universities (Government of Spain).

Use of 360-Degree Video
in Organizational Communication: Case
Study of Humanitarian Aid NGOs

Sara Pérez-Seijo and Berta García-Orosa

Abstract Virtual reality and 360-degree video are being used in several areas. Journalism is among them. In fact, the use of these technologies and formats has given way to a novel trend known as Immersive Journalism. But organizational communication and, more specifically, humanitarian aid NGOs have found in 360 video storytelling an opportunity to bring the realities where they work closer to the society. The aim of this immersive narrative is to allow users to become witnesses through a first-person experience of the stories' events. And connected to this, to promote the creation of ties with 'the others' and their realities. In this study, it is analyzed from a critical and ethical point of view the 360-degree video content produced by five well-known NGOs: International Federation of Red Cross and Red Crescent Societies, Doctors Without Borders, UNHCR, Save the Children and World Vision International.

Keywords Organizational communication · 360 video journalism · Immersive journalism · Public relations

1 Introduction

The relationship between journalism and organizational communication has been the subject of debate in academia and in the profession due to the mutual and continuous incursions in the different phases of information production that register a co-evolving development (Löffelholz 2004) that turned them into interdependent systems (Grossenbacher 1986). Schönhagen and Meißner (2016) point out how in the 1980s the relationship between public relations and journalism aroused interest in the Communication Sciences, but as the influence of Public Relations had already been discussed in the 1920s at the 7th Conference of German Sociologists in 1930 in Berlin, and in 1866 Wuttke had already lamented the great influence in journalistic

S. Pérez-Seijo (✉) · B. García-Orosa
Universidade de Santiago de Compostela, Santiago de Compostela, Spain
e-mail: s.perez.seijo@usc.es

B. García-Orosa
e-mail: berta.garcia@usc.es

© Springer Nature Switzerland AG 2020
J. Vázquez-Herrero et al. (eds.), *Journalistic Metamorphosis*,
Studies in Big Data 70, https://doi.org/10.1007/978-3-030-36315-4_8

reporting of what were known as press offices (*Pressbüros*), especially in political parties.

Since then, the interrelation and influence of the organizational communication, understood as departments for the management of communication between the organization and its public, went through the press agent model with asymmetrical and unidirectional communication up to the dircom and his influence on the media agenda have been continuous. Although there are no conclusive studies on its incidence and its determining factors, the production of journalistic information today is hardly imaginable without the support of institutional sources.

On this journey, the arrival of digital communication represents a new stage in this evolution for several reasons: citizens' access to sources directly, sources of information 2.0, influence of digital organizational communication and advances and technological innovations and adaptation of organizational communication to the network. The ongoing debate lies between the possibilities of democratizing information by Internet and the homogenization of access to it due to the greater influence of the organizational communication.

During the last 25 years, journalism has undergone changes in all phases of the productive system, combining old and new media (Jenkins 2003) and building a hybrid system of media and journalistic practices (Chadwick 2013; Hamilton 2016) that seeks formulas to combine immediacy with depth and obtain the citizen engagement, above all in the form of permanence on the page and, therefore, economic benefits.

This is one of the objectives, as will also be explained throughout the chapter. The cabinets also interfered in the digital world from the beginning, using the Internet as a repository for documentation and giving rise to the so-called 1.0 cabinets. Over the years, the projects have been modified, seeking above all interactivity and, at the final stage, the citizen engagement. From the beginning with the web and the e-mail with a unidirectional and asymmetric communication being cabinets 1.0; in a second stage with the blogs (2004), Facebook and YouTube (2004–2006), and Twitter (2007) and a last one with the search of new narratives and, as it happens with journalism, in a hybrid communication.

Within this latest phase, one of the emerging, most relevant and least studied innovations are the new narratives (Brubaker et al. 2018; Fraustino et al. 2018) as an outstanding element for influencing different audiences, but especially in the media. New technological trends on the Internet (Serrano-Cobos 2016) and an important commitment to innovation and experimentation (Peñafiel 2016; Bender 2004) have opened the journalism and organizational communication to digital narrative. Initially migrated (mainly textual), they evolved from an initial adaptation to the characteristics of the Internet (Bolter 1991), towards narratives conceived as digital and multiform with incipient characteristics and analysis approaches.

The search for new ways of counting tries to attract the attention of the receiver and, above all, to get his commitment and this has become a key element in online organizational communication (López-Hermida-Russo and Vargas-Monardes 2013; Maarek 2014; D'Adamo and García-Beaudox 2016). Its definition is not yet univocal (García-Orosa et al. 2017) although important advances have been made since

Todorov (1969) coined the term 'narratology' to designate the new theory of literary narration (Smith 1981; Phelan 2005; Ochs and Capps 2001; Seale 2000) which gave rise to different models of analysis. With these antecedents, scientific literature approached narratives from three important moments: the primacy of the text as a documentary source; narrative is approached as a specific textual structure; it is observed as a polymorphic phenomenon determined by its communicative context. In this last phase, it is understood that narration goes beyond traditional limits of narratology and fiction and it is conceived as a valid theoretical framework to confront the transformations and characteristics of digital narrative in political parties, the narration is assumed as a representation of the experience constructed through discourse in which meaning is given, communication is made possible and action is oriented (Hyvärinen 2008). From this concept of digital narrative, we will approach the subsequent analysis, but also bearing in mind that language is the hybridization processes (Chadwick 2013; Hamilton 2016) and the search for new narratives (Jenkins 2003; Shin and Biocca 2017). Today, digital language is multichannel, polysynthetic and is integrated within a narrative with hyper-fragmented textualities and with narrative programmes in which the combination of different elements and actors participating in the elaboration of the messages is sought (Herring and Androutsopoulos 2015; Adami 2016).

Different conceptualizations of these changes have been established which, except the last, form part of a broad semantic family: crossmedia, multiplatform, multimodality, hyperseriality, hyperdiegesis, hybrid media, transformation, transmedial narratives, multiple platforms and intertextual merchandise (Rodríguez-Ferrándiz and Peñamarín 2014; Guerrero-Pico and Scolari 2016) and immersive (De la Peña et al. 2010; Domínguez 2013), in which this article will focus due to its characteristics and special relevance in the relationship between journalism and organizational communication.

This innovation acquires a special relevance in public relations in NGOs due to its need for interaction and, above all, commitment to the audiences that are the core of its existence. In this sense, Fraustino et al. (2018) point out the increase in the use of 360-degree in NGOs and demonstrate that it has a greater impact on the public, above all due to the sense of spatial presence. Suh et al. (2018) also highlight engagement and Cummings and Bailenson (2016) refer to virtual reality and 360-degree video with two advantages and one challenge: engagement, presence, both emotional and spatial, and ethics.

1.1 360 Video Journalism and Humanitarian Communication

The use of 360-degree videos to tell the news broke the prevailing models of information consumption. For the first time, users could cross the window that the screen represented until then and to be there, within the scene. Not only in real-image, since the use of virtual reality techniques also brought new opportunities to recreate places, persons and even situations based on real evidence. In this sense, Nonny de la Peña was a pioneer in the use of such resources to create factual recreations trough computer-generated imagery, as is the case of the virtual reality film *Hunger in Los Angeles* (2012). De la Peña saw in the immersive narratives an opportunity to report social realities and to bring it closer to the users.

In 2015 the use of the 360-degree videos and the virtual reality began to spread among newsrooms from news outlets all over the world (Doyle et al. 2016), as in the case of *The New York Times*, *Chosun Ilbo*, *Russia Today*, *El País*, *The Guardian*, *Associated Press* or the *British Broadcasting Corporation*. And simultaneously, the NGOs, and humanitarian communication as a whole, also saw in these resources an opportunity to bring closer the social realities where they work (Nash 2018; Soler-Adillon and Sora 2018). In fact, in 2015 was released *Clouds Over Sidra*, a 360-degree documentary about the Za'atari refugee camp in Jordan produced by Gabo Arora and Chris Milk, and in collaboration with the United Nations and the technology company Samsung. But while this type of content production opened new opportunities to storytelling, it also brought ethical challenges (Kent 2015; Bartzen 2015; Kool 2016; Aitamurto 2018; Pérez-Seijo and López-García 2019a, b; Sánchez Laws and Utne 2019).

These resources have opened a new evolutionary stage of multimedia (Salaverría 2016) in which immersion on scene and place illusion are particularly significant (De la Peña et al. 2010). The use of virtual reality and 360-degree video to create news stories has been called Immersive Journalism. De la Peña et al. defined this new model as "the production of news in a form in which people can gain first-person experiences of the events or situation described in news stories" (2010: 291).

But we can distinguish two main modalities within this trend: 360 video journalism or 360 storytelling (Van Damme et al. 2018; Elmezeny et al. 2018); and VR storytelling or VR journalism (Sánchez Laws and Utne 2019). However, the 360-degree video is the most common format in the immersive production of news outlets (Hardee and McMahan 2017), especially the real-image ones. Behind this generalization there are several reasons, such as the economic cost—equipment and creation process, the production times and the possibility of a multi-platform dissemination (Pérez-Seijo and López-García 2018). In this respect, media and news outlets tend to publish and spread their contents through the main social platforms, such as Facebook, YouTube and even Vimeo (Sidorenko et al. 2017; Van den Broeck et al. 2017; Pérez-Seijo et al. 2018). A decision, or strategy, that ensures a greater democratization of the user's access to the spherical videos (Pérez-Seijo and López-García 2018).

Immersive Journalism came as a journalistic revolution. A new form of non-fiction content production based on three main elements: place illusion, empathy and user engagement. The first one, the feeling of being present there, on the scene, through the use of a virtual reality headset or a low-cost version such as the Google Carboards. These immersive films allow users to get a first-person experience of the news events with a limited but possible interaction: 360-degree view and, sometimes, an interactive navigation.

Secondly, some authors and professionals claim that virtual reality and 360-degree films enhance the feeling of empathy with the others and their realities (Milk 2015; Constine 2015; Kool 2016; Sánchez Laws 2017). Nevertheless, there is still not enough scientific evidence to endorse this thought (Shin 2018; Van Damme et al. 2018). Even so, this brings new possibilities to humanitarian aid communication. And thirdly, several studies suggest that immersive technologies create higher levels of audience engagement (Suh et al. 2018; Wang et al. 2018; Bindman et al. 2018; Shin and Biocca 2017).

Immersive Journalism allows users to become 'immersive witnesses' (Nash 2018) of the others suffering, although with the moral risks involved in putting themselves in the place of others and try to understand a social reality from a far removed and privileged view. Chouliaraki understands this improper distance as "practices of communication that (...) make use of imaginative textualities that problematize the act of representation itself and, thereby, privilege the voices of the West over the voices of suffering others" (2011: 365).

But to promote these aspects, professionals and journalists carry out some practices that sometimes are not based on an ethical decision-making. Pretending that sources/characters and users are face to face is one of the main strategies that news outlets and virtual reality producers use to strengthen the link between viewer and story. And this, together with the re-enactment of the sources/characters actions and even the staging, has opened an ethical discussion on what the limits should be and for what purpose (Kool 2016; Kent 2015; Bartzen 2015; Aitamurto 2018; Pérez-Seijo and López-García 2019a, b; Sánchez Laws and Utne 2019).

2 Methodology

The aim of this study is to gain an in-depth understanding about the use of the 360-degree video by NGOs. In particular, how these entities produce immersive experiences that act as awareness campaigns for the general public. However, the purpose of this document is twofold: on the one hand, to observe and to compare the formal features of these type of contents; and on the other hand, to reflect on the ethical challenges that these videos put on the table.

To this end, we have combined a literature review on the topic—directly or indirectly related to this—with a comparative analysis of cases, specifically five 360-degree videos produced by five of the major humanitarian aid NGOs: *Rescuing people in the Mediterranean*, by the International Federation of Red Cross and Red Crescent Societies (IFRC); *We fled a war, then we nearly drowned*, by Doctors Without Borders (MSF); *Step inside a Rohingya tent Kutupalong refugee camp, Bangladesh*, by UNHCR; *On board our life-saving ship*, by Save the Children; and *Ali's story* by World Vision International (WVI).

This represents a diverse sample, but our goal was to get an overview of how these NGOs apply the immersive storytelling on their contents. We have selected the five above mentioned videos because they are real-image, and they address the

Table 1 Data sheet of the five selected cases

Video	Rescuing people in the Mediterranean	We fled a war, then we nearly drowned	Step inside a Rohingya tent Kutupalong refugee camp	On board our life-saving ship	Ali's story
NGO	IFRC	MSF	UNHCR	Save the Children	WVI
Year	2016	2016	2017	2016	2016
Form/genre	Simple video	Report	Report	Report	Documentary
Duration	0:04:07	0:01:39	0:01:17	0:00:54	0:05:32
Language	–	English (subtitle)	English	English	English
Image	Real	Real	Real	Real	Real
Topic	Migrant rescue in the Mediterranean sea	Refugees in a Greek camp	Life of Rohingya refugees in the Kutupalong camp	Migrants rescue operations at the sea	Life of a family in a Lebanese refugee camp
YouTube channel	IFRC	Médecins Sans Frontières/Doctors Without Borders	UNHCR, the UN Refugee Agency	SaveTheChildren	World Vision International

Own elaboration

same topic from different angles: the migrant crisis. Furthermore, all these pieces are available on YouTube—in the NGOs principal accounts, so users can watch them through a mobile device, a desktop computer or even a virtual reality headset. This last option is important as it is considered the most immersive form of consumption (Table 1).

3 Results

3.1 Formal Features

As mentioned above, this study is based on the comparative analysis of five 360-degree videos produced by five of the most influential humanitarian aid NGOs. These contents present several differences regarding its conceiving and purpose, but we have also found formal likeness and similar approaches.

In this regard, the aim of each immersive video is the first common feature that it is possible to observe. Although we selected these five pieces because they address migrant issues, we have noted that the main objective of the NGOs was to reflect the actual reality of these people. However, each spherical video allows a distinct perspective, as explained below.

- *Rescuing people in the Mediterranean* (IFRC): users witness first-hand how a rescue operation is carried out by an NGO ship.
- *We fled a war, then we nearly drowned* (MSF): a video about the life in a Greek refugee camp.
- *Step inside a Rohingya tent Kutupalong refugee camp, Bangladesh* (UNHCR): as anticipated in the title, the purpose of this immersive experience is to raise awareness about the reality in a Rohingya refugee camp located in Bangladesh.
- *On board our life-saving ship* (Save the Children): diverse images of a migrant rescue at sea accompanied by an invitation to donate money.
- *Ali's story* (WVI): the video is about the life of Ali, a thirteen-years-old child, and his family in a Lebanese refugee camp.

All these pieces are real-image videos, so we have not found any scene produced through computer-generated imagery (CGI) nor even hybrid. Nevertheless, there are significant differences with regard to the form. While we have described *Rescuing people in the Mediterranean* as a simple video—later we will explain why, the remaining videos are more elaborate. On the one hand, *Ali's story* is an immersive documentary. On the other hand, *We fled a war, then we nearly drowned*, *Step inside a Rohingya tent Kutupalong refugee camp, Bangladesh* and *Ali's story* are 360-degree video reports.

Except in *Rescuing people in the Mediterranean*, all the videos include a narrator. In the case of *Ali's story*, a voice-over of a female journalist guide users through the documentary as well as two human sources give their own testimonies. In contrast, the role of storytellers is assumed by a human source in the reports *We fled a war, then we nearly drowned* (male voice-over and present on the scene) and *Step inside a Rohingya tent Kutupalong refugee camp, Bangladesh* (the UNHCR Emergency Response Coordinator explains the situation there). And in the case of *On board our life-saving ship*, the narrator is a text instead a human voice.

It should be noted that despite the issues addressed, none of the videos contains graphic or sensitive content. This represents an important finding because we also wanted to see whether the producers had included images of that kind in pursuit of a more powerful effect. As mentioned in a previous section, some authors attribute to 360-degree videos the possibility of experience higher levels of empathy although there is yet no significant evidence of that. However, this has brought new ethical challenges as some practices and procedures carried out by journalists and professionals are not based on an ethical decision-making (Pérez-Seijo and López-García 2019a, b).

3.2 Postproduction and Staging

One of the main challenges of the Immersive Journalism is the image integrity. As place illusion is the main goal of these experiences, journalists and professionals can be attempted to take decisions that go against the existing ethical guidelines. As already mentioned, digital manipulation and staging are among these practices. The first technique is often used to remove the tripod from the image to not interfere with the possible user's feeling of place illusion. The second one, to pretend that human sources/characters are face to face with the users, as if the viewers were really there, on the scene.

Regarding the erasing of the camera support, we have found this situation in two of the five videos analysed. Specifically, the tripod disappears in the report *On board our life-saving ship* and in the documentary *Ali's story*. No unified criteria were observed in *We fled a war, then we nearly drowned*, as in some scenes it is visible and in others it has completely vanished. By contrast, in *Rescuing people in the Mediterranean* and in *Step inside a Rohingya tent Kutupalong refugee camp, Bangladesh* the support has been replaced by an image, a black circle and the logo of the Red Cross and Red Crescent Societies respectively. On the contrary, it is also possible to observe some superimposed elements on the images of all videos, but none of these has further impact as its function is to hide the camera support (circle and logo) or to add basic information (location, title and NGO logo).

On the other hand, no visual or sound effects have been found in these immersive videos. However, two of them include extradiegetic music from the begging to the end: *On board our life-saving ship* and *Ali's story*. The music is instrumental in both cases. Nevertheless, manifold ethical codes warn about the power of music to introduce subjectivity and therefore biases in the stories. In this sense, The Radio Television Digital News Association (2015) states in its guidelines for *Ethical Video and Audio Editing* that

> Music, especially, has the ability to send complex and profound editorial messages. [...] However, if the music is a soundtrack audio recording, then journalists must ask themselves whether the music adds an editorial tone to the story that would not be present without the music. (The Radio Television Digital News Association 2015)

So, the question here is if humanitarian communication should be considered as journalism or as a marketing strategy for social purposes. In fact, *On board our life-saving ship* is a clear invitation from Save the Children to donate. On the other hand, UNHCR also includes a request to donate in the YouTube's description of the video *Step inside a Rohingya tent Kutupalong refugee camp*.

It should be noted that there is a clear staging in *Step inside a Rohingya tent Kutupalong refugee camp, Bangladesh*. Users are face to face with the source when he is on the scene, in others we can only hear him in voice-over. It seems the man is talking to us as he and those persons accompanying them also look at us, although they are actually staring at the camera. However, if users try to figure out what is behind their backs, they will find a journalist who at certain moments gives indications to the source for staring at the camera, even he shows him a written paper.

In *Step inside a Rohingya tent Kutupalong refugee camp, Bangladesh* is Joung-ah Ghedini-Williams, UNHCR Emergency Response Coordinator, who carries the camera while recording. She explains the situation of the newcomers by also addressing us ("as you can see"). Perhaps this is not staging as such.

3.3 Users on the Scene

As the aim of the Immersive Journalism is to create place illusion, we also wanted to observe how users are represented on scene. In these particular cases, they access to the narrative world (the news world or reality) as themselves, and not as a character or an avatar. But, obviously, their bodies are invisible on the scene. Despite this invisibility, users are noticed—although it is pretended—by the characters or sources of the story (face to face, gazes or even allusions) in *Step inside a Rohingya tent Kutupalong refugee camp, Bangladesh* and *On board our life-saving ship*. This simulation allows viewers to leave their passive role as observers and to become witnesses of the realities. Users not only gain a first-person view through a virtual reality headset, but also could experience place illusion and feel that characters or sources address him or her, as in the aforementioned videos.

This opens new opportunities for humanitarian communication since the public can immerse themselves in the places where NGOs work, and therefore know first-hand the diverse social realities and contexts. From being at home to being in a refugee camp in Bangladesh thanks to the use of a virtual reality headset.

On the other hand, users do not maintain a steady height throughout the video. In the case of *Rescuing people in the Mediterranean* and in *Step inside a Rohingya tent Kutupalong refugee camp, Bangladesh* a person is holding the camera, so the perspective adopted by the user is strange. In the remaining videos, the height of the tripod changes in some scenes. In *On board our life-saving ship* it is because a camera person is holding a long tripod and its length varies at certain moments of the video, but which calls to mind a high-angle shot. Finally, in *Ali's story* and in *We fled a war, then we nearly drowned* we found a combination of 'shots' and heights.

4 Conclusions

NGOs may have found a narrative that meets three of the digital communication requirements: to have a democratized access and consumption; to promote user engagement; and to be innovative. 360 video journalism is a novel storytelling form that allows NGOs to produce content that draw audience attention to their stories on specific social realities and also encourage the engagement of the viewers. And this represents the key of the organizational communication in the online sphere.

Through this study, we have noted that the mission of NGOs is to promote social awareness by taking advantage of the features and possibilities of the 360-degree video. Thanks to the use of a virtual reality headset, users can go from being in their living room to being in a Lebanese refugee camp. This immersive storytelling allows NGOs to bring closer the realities where they help to the society.

But this closeness poses also ethical challenges that NGOs in particular, and organizational communication in general, should deal with. The introduction of possible bias in pursuit of a specific reaction or emotion is one of these. Although this is not a new conflict, it is compounded by the power of the immersion in 360-degree video. A possibility that some authors has connected to a more empathic experience, even if there is yet no convincing evidence of that. And this is where an important question emerges: Where are the limits between reporting and marketing? Does one lead to the other? In the documentary *Ali's Story*, World Vision International first tells the story of Ali and his family and then the NGO takes the opportunity to emphasize the importance and need for its schools and services to children like Ali. Obviously, NGO communication is intended to generate a social impact and to obtain an economic return, as it is fundamental to do their work on the ground.

The biggest benefit of 360-degree videos is allowing users to experience first-hand the social realities, to turn them into witnesses on the scene thanks to the use of a virtual reality headset. However, this immersive storytelling form also can serve to put users in the shoes of a particular person in a specific situation and therefore see the world through his or her eyes. But behind these possibilities there is a moral risk: to create improper distance and to convey a Western and privileged vision.

In short, NGOs have a dual responsibility. On the one hand, to not misrepresent the social realities to make a greater impact. An ethical decision-making is crucial to respect the affected people. And on the other hand, to use the possibilities offered by novel narratives to communicate better the messages, not to seek a direct and specific empathetic reaction. Here is where the ethical boundaries arise.

Acknowledgements This research has been developed within the project *Digital native media in Spain: storytelling formats and mobile strategy* (RTI2018-093346-B-C33), from the Ministry of Science, Innovation and Universities (Government of Spain) and co-funded by the ERDF structural fund. Furthermore, this study has been developed within the framework of the activities of the Novos Medios Research Group (GI-1641) from the Universidade de Santiago de Compostela, supported by the Program of Consolidation and Structuring of Competitive Research Units of Regional Government of Galicia—Xunta de Galicia—(ED431B 2017/48). On the other hand, the author Sara Pérez-Seijo is beneficiary of the Training University Lecturers' (FPU) Program funded by Spanish Ministry of Science, Innovation and Universities (Spanish Government).

References

Adami E (2016) Multimodality. In: García O, Flores N, Spotti M (eds) The Oxford handbook of language and society. Oxford Handbook Online, Oxford, pp 451–473
Aitamurto T (2018) Normative paradoxes in 360 journalism: contested accuracy and objectivity. New Media Soc 21(1):3–19. https://doi.org/10.1177/1461444818785153
Bartzen K (2015) Virtual journalism: immersive approaches pose new questions. Center for Journalism Ethics. Retrieved from https://goo.gl/9937LN

Bender W (2004) The seven secrets of the media lab. BT Technol J 22(4):5–6. https://doi.org/10.
    1023/B:BTTJ.0000047629.12018.33
Bindman SW, Castaneda LM, Scanlon M, Cechony A (2018) Am I a bunny?: the impact of high and
    low immersion platforms and viewers' perceptions of role on presence, narrative engagement,
    and empathy during an animated 360° video. In: Proceedings of the 2018 CHI conference on
    human factors in computing systems. ACM, New York
Bolter JD (1991) Writing space. The computer, hypertext, and the history of writing. Lawrence
    Erlbaum Associates, New Jersey
Brubaker P, Church S, Hansen J, Pelham S, Ostler A (2018) One does not simply meme about
    organizations: exploring the content creation strategies of user-generated memes on Imgur. Public
    Relat Rev 44(5):741–751. https://doi.org/10.1016/j.pubrev.2008.06.004
Chadwick A (2013) The hybrid media system: politics and power. Oxford University Press, Oxford
Chouliaraki L (2011) 'Improper distance': towards a critical account of solidarity as irony. Int J
    Cult Stud 14(4):363–381
Constine J (2015) Virtual reality, the empathy machine. TechCrunch. Retrieved from http://goo.gl/
    VYOK1w
Cummings JJ, Bailenson JN (2016) How immersive is enough? A meta-analysis of the effect of
    immersive technology on user presence. Media Psychol 19(2):272–309
D'Adamo O, García-Beaudox V (2016) Comunicación política: narración de historias, construcción
    de relatos políticos y persuasion. Comun Hombre 12:23–39
De la Peña N, Weil P, Llobera J, Giannopoulos E, Pomés A, Spanlang B, Friedman D, Sánchez-
    Vives M, Slater M (2010) Immersive journalism: immersive virtual reality for the first-person
    experience of news. Presence: Teleoperators Virtual Environ 19(4):291–301. https://doi.org/10.
    1162/pres_a_00005
Domínguez E (2013) Periodismo inmersivo: La influencia de la realidad virtual y del videojuego
    en los contenidos informativos. Editorial UOC, Barcelona
Doyle P, Gelman M, Gill S (2016) Viewing the future? Virtual reality in journalism. Retrieved from
    https://goo.gl/ZJX4UG
Elmezeny A, Edenhofer N, Wimmer J (2018) Immersive storytelling in 360-degree videos: an
    analysis of interplay between narrative and technical immersion. J Virtual Worlds Res 11(1)
Emblematic Group (2012) Hunger in L.A. Retrieved from https://emblematicgroup.com/
    experiences/hunger-in-la/
Fraustino J, Lee JY, Lee SY, Ahn H (2018) Effects of 360° video on attitudes toward disaster
    communication: mediating and moderating roles of spatial presence and prior disaster media
    involvement. Public Relat Rev 44(3):331–341. https://doi.org/10.1016/j.pubrev.2018.02.003
García-Orosa B, Vázquez-Sande P, López-García X (2017) Narrativas digitales de los principales
    partidos políticos de España, Francia, Portugal y Estados Unidos. Prof Inf 26(4):589–600. https://
    doi.org/10.3145/epi.2017.jul.03
Grossenbacher R (1986) Hat die "Vierte Gewalt" ausgedient? Zur Beziehung zwischen Public
    Relations und Medien [Is the 'fourth estate' disused? On the relationship between PR and media].
    Media Perspekt 17:725–731
Guerrero-Pico M, Scolari C (2016) Narrativas transmedia y contenidos generados por los usuarios:
    el caso de los crossovers. Cuadernos.info 8:183–200. https://doi.org/10.7764/cdi.38.760
Hamilton J (2016) Hybrid news practices. In: Witschge T, Anderson C, Domingo D, Hermida A
    (eds) The SAGE handbook of digital journalism. Sage, London, pp 164–178
Hardee GM, McMahan R (2017) FIJI: a framework for the immersion-journalism intersection.
    Front ICT 4(21):1–18
Herring S, Androutsopoulos J (2015) Computer-mediated discourse 2.0. In: Tannen D, Hamilton
    HE, Schiffrin D (eds) The handbook of discourse analysis, 2nd edn. Wiley, Chichester, pp 127–151
Hyvärinen M (2008) Analyzing narratives and storytelling. In: Alasuutari P, Bickman L, Brannen
    J (eds) The SAGE handbook of social research methods. Sage, London, pp 447–460. https://doi.
    org/10.4135/9781446212165.n26

Jenkins H (2003) Transmedia storytelling. Moving characters from books to films to video games can make them stronger and more compelling. MIT Technology Review. Retrieved from https://www.technologyreview.com/s/401760/transmedia-storytelling/

Kent T (2015) An ethical reality check for virtual reality journalism. Medium. Retrieved from https://goo.gl/DQUe1Q

Kool H (2016) The ethics of immersive journalism: a rhetorical analysis of news storytelling with virtual reality technology. Intersect 9(3):1–11

Löffelholz M (2004) Ein privilegiertes Verhältnis. In: Löffelholz M (ed) Theorien des Journalismus. VS Verlag für Sozialwissenschaften, Wiesbaden

López-Hermida-Russo AP, Vargas-Monardes J (2013) La política relatada: el storytelling de Barack Obama en el marco de la Operación Gerónimo. Palabra Clave 16(1):12–44. https://doi.org/10.5294/pacla.2013.16.1.1

Maarek P (2014) Politics 2.0: new forms of digital political marketing and political communication. Trípodos 34:13–22

Milk C (2015) How virtual reality can create the ultimate empathy machine. In: Ted talks 2015, Vancouver. Retrieved from https://goo.gl/ZW1Nj9

Nash K (2018) Virtual reality witness: exploring the ethics of mediated presence. Stud Doc Film 12(2):119–131. https://doi.org/10.1080/17503280.2017.1340796

Ochs E, Capps L (2001) Living narrative. Creating lives in everyday storytelling. Harvard University Press, Cambridge

Peñafiel C (2016) Reinvención del periodismo en el ecosistema digital y narrativas transmedia. AdComunica 12:163–183. https://doi.org/10.6035/2174-0992.2016.12.10

Pérez-Seijo S, López-García X (2018) Las dos caras del Periodismo Inmersivo: el desafío de la participación y los problemas éticos. In: López Paredes M (ed) Nuevos escenarios en la comunicación: retos y convergencias. Centro de Publicaciones PUCE, Quito, pp 279–305

Pérez-Seijo S, López-García X (2019a) Five ethical challenges of immersive journalism: a proposal of good practices' indicators. In: Rocha Á, Ferrás C, Paredes M (eds) Information technology and systems. ICITS 2019. Advances in intelligent systems and computing, vol 918. Springer, Cham, pp 954–964

Pérez-Seijo S, López-García X (2019b) La ética del Periodismo Inmersivo a debate. Hipertext.net 18:1–13. https://doi.org/10.31009/hipertext.net.2019.i18.01

Pérez-Seijo S, Melle Goyanes M, Paniagua Rojano FJ (2018) Innovación en radiotelevisiones públicas europeas: narrativas inmersivas y organización de los contenidos 360 grados en plataformas digitales. Rev Lat Comun Soc 73:1115–1136

Phelan J (2005) Living to tell about it. A rhetoric and ethics of character narration. Cornell University Press, Ithaca

Radio Television Digital News Association (2015) Code of ethics. Retrieved from https://www.rtdna.org/content/rtdna_code_of_ethics

Rodríguez-Ferrándiz R, Peñamarín C (2014) Narraciones transmedia y construcción de los asuntos públicos. CIC—Cuad Inf Comun 19:9–16

Salaverría R (2016) Los medios de comunicación que vienen. In: Sádaba C, Martínez-Costa MP, García JA (eds) Innovación y desarrollo de los cibermedios en España. EUNSA, Pamplona, pp 255–263

Sánchez Laws AL (2017) Can immersive journalism enhance empathy? Digit J [Online first, 20 Oct]

Sánchez Laws AL, Utne T (2019) Ethics guidelines for immersive journalism. Front Robot AI 6(28):1–13

Schönhagen P, Meißner M (2016) The co-evolution of public relations and journalism: a first contribution to its systematic review. Public Relat Rev 42(5):748–758

Seale C (2000) Resurrective practice and narrative. In: Andrews M, Sclater SD, Squire C, Treacher A (eds) Lines of narrative. Psychosocial perspectives. Routledge, London-New York

Serrano-Cobos J (2016) Tendencias tecnológicas en internet: hacia un cambio de paradigma. Prof Inf 25(6):843–850. https://doi.org/10.3145/epi.2016.nov.01

Shin D (2018) Empathy and embodied experience in virtual environment: to what extent can virtual reality stimulate empathy and embodied experience? Comput Hum Behav 78:64–73

Shin D, Biocca F (2017) Exploring immersive experience in journalism. New Media Soc 20(8):1–24. https://doi.org/10.1177/1461444817733133

Sidorenko P, Cantero de Julián JI, Herranz de la Casa JM (2017) La realidad virtual y el formato multimedia en 360° como mecanismo de enriquecimiento de los contenidos periodísticos. In: Sierra J (ed) Nuevas tecnologías audiovisuales para nuevas narrativas interactivas digitales en la era multidispositivo. McGraw Hill, Madrid, pp 99–108

Smith BH (1981) Narrative version, and narrative theories. In: Mitchell WJT (ed) On narrative. University of Chicago Press, Chicago

Soler-Adillon J, Sora C (2018) Immersive journalism and virtual reality. In: Pérez-Montoro M (ed) Interaction in digital news media. From principles to practice. Palgrave Macmillan, Cham, pp 55–83. https://doi.org/10.1007/978-3-319-96253-5

Suh A, Wang G, Gu W, Wagner C (2018) Enhancing audience engagement through immersive 360-degree videos: an experimental study. In: International conference on augmented cognition. Springer, Cham, pp 425–443

Todorov T (1969) Grammaire du Decameron. Mouton, La Haya

Van Damme K, All A, De Marez L, Van Leuven S (2018) 360° video journalism: experimental study on the effect of immersion on news experience and distant suffering. J Stud [Online first, 3 Jan]

Van den Broeck M, Kawsar F, Schöning J (2017) It's all around you: exploring 360° video viewing experiences on mobile devices. In: Proceedings of the 2017 ACM on multimedia conference. ACM, Mountain View, CA, pp 762–768

Wang G, Gu W, Suh A (2018) The effects of 360-degree VR videos on audience engagement: evidence from the New York Times. In: International conference on HCI in business, government, and organizations. Springer, Cham, pp 217–235

Sara Pérez-Seijo Ph.D. Student in Communication and Contemporary Information from Universidade de Santiago de Compostela (USC) and member of the Novos Medios research group at the same university. She is beneficiary of the Training University Lecturers' (FPU) Programme funded by Spanish Ministry of Science, Innovation and Universities (Spanish Government). Her research is linked to non-fiction digital storytelling, focusing on the VR and 360° video storytelling and the immersive narratives as a whole.

Berta García-Orosa Professor of Journalism at Universidade de Santiago de Compostela. She holds a degree in Journalism and in Political Science and Administration, also a Ph.D. in Journalism. Her main research interests are digital media and organizational and political communication. Her latest publications include *Language in social networks as a communication strategy: public administration, political parties and civil society* and *Algorithmic communication and political parties: Automation of production and flow of messages*.

Shared Spaces for News Content Production in Spanish Online Media

Ainara Larrondo Ureta, Koldo Meso Ayerdi and Simón Peña Fernández

Abstract This chapter explores three dynamics underlying the evolution of the professional, entrepreneurial and structural aspects of the digital ecosystem in which journalism now takes place: direct interaction between the media and audiences, an expanding range of formats for disseminating content and new opportunities for branding. Analysis covers professional practices and profiles, business models, cross-branding strategies and mobile audience engagement and impact in the Spanish media market. The sample employed for this study, composed of seven media of record that collectively represent a broad spectrum of media and communication groups active in Spain today, includes two newspapers offering both print and online editions (*El País* and *El Mundo*, published respectively by the Prisa Group and Unidad Editorial), one regional newspaper (*Diario de Navarra*), the Spanish public broadcasting company *RTVE* and three digital native news enterprises that publish exclusively online editions (*El Diario*, *El Español* and *El Huffington Post*).

Keywords Innovation · Media · Strategy · Spain

1 Introduction

Digitalization and the application of new technologies have transformed the way in which news is consumed and triggered a structural reorganization of the manner in which news content is produced.

News enterprises accustomed to focusing on a single medium have been obliged to alter their internal and external operational frameworks and pursue multiplatform models that require adapting content for delivery in multiple formats (Salaverría

A. Larrondo Ureta (✉) · K. Meso Ayerdi · S. Peña Fernández
University of the Basque Country, Leioa, Spain
e-mail: ainara.larrondo@ehu.eus

K. Meso Ayerdi
e-mail: koldo.meso@ehu.eus

S. Peña Fernández
e-mail: simon.pena@ehu.eus

© Springer Nature Switzerland AG 2020
J. Vázquez-Herrero et al. (eds.), *Journalistic Metamorphosis*,
Studies in Big Data 70, https://doi.org/10.1007/978-3-030-36315-4_9

2010). As a consequence, the structures and processes that characterized journalism throughout the twentieth century have been gradually supplanted by others linked to the requisites and possibilities of the digital twenty-first century world (Salaverría and Negredo 2008). While journalists' functions and competences may have changed substantially during this period, public expectations concerning journalism's role in society are as high (or higher) than ever, a circumstance that is making the work of professional communicators more difficult (McQuail 2013).

In a media landscape made considerably more challenging by the ubiquitous presence of technology (Rondón and Leyva 2017), digital convergence has upped the ante by facilitating the integration of tools, spaces, working methods and languages formerly pertaining to distinct and separate operations (López García and Pereira 2010)—a multi-faceted process that has irrevocably affected the technological, business, professional and editorial aspects of journalism. There are nevertheless a variety of factors currently conditioning or hindering fuller convergence in all of these areas, the most patent of which are constrained operating budgets, a difficult-to-bridge generation gap in newsrooms, incompatibilities between conventional and online content production processes, a dearth of specialized professionals in key areas, a lack of entrepreneurial vision and integrated strategies, a shortage of funds for implementation and the slow pace of technological renovation in the sector (Killebrew 2005; Dupagne and Garrison 2006; Larrondo et al. 2016).

Individuals engaged in emerging fields such as data journalism and immersive and transmedia journalism are expected to perform a broad spectrum of tasks (López García, Rodríguez and Pereira 2017). This has supposed a need for a new approach to professional training (Lewis and Westlund 2016) that strikes the right balance between traditional competences such as writing and technical skills (Deuze 2017).

Multi-channel strategies pursued by Spain's leading news outlets involving the dissemination of cross-media and transmedia content via social media platforms have begun to pay off from the perspective of multiplatform audience engagement. Cross-media strategies, the objective of which is to disseminate given content via as many platforms as possible, are admittedly less sophisticated and ambitious than transmedia strategies, which call for parcelling out a story in bits through the channels best suited for conveying specific angles and if properly designed generate much higher levels of user interaction and engagement and favor the virilization of content.

A number of scholars have noted that most news organizations in Spain distribute cross-media and transmedia content via social media to "improve their relationships with users but do not explore into other aspects of social media communications such as production or the differentiated management of platforms" (García Orosa and López García 2016: 127). This study has focused on determining the degree to which media outlets of reference in Spain have implemented cross-media and transmedia strategies for social networking platforms and gaining a clearer understanding of the most important challenges they face in terms of the management of related in-house tasks, content production and the development of formats specifically for these sites and multiplatform promotion.

News organizations have begun to restructure their editorial and production processes so as to move in this direction and innovate to the extent that their particular characteristics and resources allow. This study posits that the implementation of multi-channel strategies has prompted media enterprises in Spain to rethink their branding and external communications processes (Tosoni et al. 2017). They have also attempted to differentiate between these organizations' incursions into cross-media and transmedia branding, the latter of which has yet to be widely exploited by the sector but has nevertheless sparked growing interest among news professionals and media scholars (Siegert et al. 2015).

2   Methodology

This chapter provides an overview of strategies pursued by a representative spectrum of mainstream players in the Spanish news sector. Representativity has been determined on the basis of a combination of standard objective criteria employed by the *Oficina de Justificación de la Difusión* (OJD), the *Estudio General de Medios* (EGM) and Comscore to measure audience share and engagement and newspaper circulation in Spain and further, more specific, criteria used to ascertain the relevance of individual media organizations to the objectives of this study that took factors such as innovation, organizational and operational characteristics and business models into account. The sample compiled by means of this process was composed of *Diario de Navarra*, *El Diario*, *El Español*, *El Mundo*, *El País*, the Spanish version of *The Huffington Post* and *Radio Televisión Española* (*RTVE*). In-depth interviews were conducted with high-level representatives of all seven organizations.

The general objective of this research has been to gain a deeper understanding of the visions and strategies presently being pursued by Spanish media outlets. To this end, the authors examined the activities of a variety of sector enterprises, some of which have taken initial or tentative steps toward newsroom integration and employed basic multi-channel social media strategies to promote conventional forms of journalism and others of which have fully integrated their newsrooms, developed multimedia job profiles and implemented coordinated strategies for producing and disseminating cross-media and transmedia content.

Researchers also defined following interrelated specific objectives:

- analyze the latest trends in online news consumption;
- determine the processes, routines and dynamics related to news organizations' internal content creation strategies;
- ascertain the purposes for which news organizations use social media channels (branding, audience loyalty schemes, redirecting traffic to a corporate website, etc.); and
- determine how organizations developed multiplatform and transmedia strategies for disseminating news content and promoting their brands in the light of conditioning factors such as individual corporate visions, media category and available resources.

3   Mobile Ecosystem, Active Audiences

The introduction of Web 2.0 a decade after the emergence of online media, the expansion of mobile phone use shortly after and the game-changing debut of the iPhone in 2007 have progressively transformed the ways in which news content is produced and consumed. Since that time, mobility and interactivity have shaped the contours and parameters of media consumption (Masip et al. 2015) by allowing users not only to choose precisely what the content they consume, but how and where they consume it as well.

The proliferation of mobile devices (smartphones and tablets) has opened a new channel for disseminating content that far from serving as a merely complement to traditional media has become the main source of traffic to their online sites (Peña et al. 2016).

Veteran media outlets have managed to satisfy their audiences by promoting inter-activity and user participation that for the most part takes place on the profile pages they have established on social networking platforms, which have allowed them to boost their visibility exponentially and explore new ways of disseminating news and entertainment content to readers and viewers.

The upshot has nevertheless been the coalescence of a new set of circumstances under which conventional media have lost their monopoly of symbolic power in the news market and ceded a significant portion of the space they once controlled to brash new upstarts in an increasingly complex ecosystem. News distribution has undergone a sudden, radical decentralization (Casero Ripollés and López Meri 2015) marked by a shift from a culture anchored in ownership to a culture focused on access. Modern technology has made it possible for almost anyone to produce and disseminate content, much of which is low quality but has broad public appeal (Ufarte et al. 2018).

A breakdown of traffic sources for each of the news organizations in the sample confirms the strong shift in news consumption habits underway in Spain.

As can be observed in Fig. 1, social media networks currently generate 8.3% of the total traffic to the online news sources analyzed, which has, in any case, fallen considerably due to a series of recent changes in Facebook algorithms. Direct access, which is responsible for 45.2% of incoming traffic, and online search (normally understood as searches using the name of the news source as a key search term), which is responsible for another 39.1%, continue to be the primary modes of user access. These figures indicate that roughly one out of three visitors to the online news sources analyzed for this study accessed their sites indirectly by means of either social networking platforms or general queries via search engines. Newsrooms pursue strategies for capturing as much of this indirect traffic as possible. As a representative of *El Mundo* noted during an interview conducted for this study:

> SEO and social media traffic are very important. They may be more fickle than other types of traffic but are crucial if you're looking to be a leader and sell advertising at a higher price than your competitors, which is what [sector] leadership hinges on. (interviewee *El Mundo*, pers. comm., March 2019)

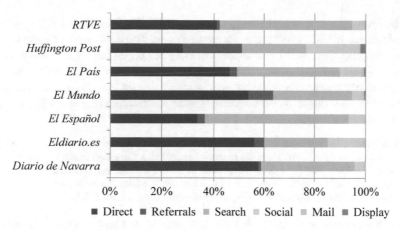

Fig. 1 Traffic sources. Similarweb

The primary traffic sources of the news media examined here vary. Whereas brand identity constitutes a major pull factor for conventional, well-established outlets newer, digital native competitors have been notably more adept at connecting with audiences via social networking platforms. Facebook generates 52.4% of the traffic to digital native outlets and Twitter (which in the case of their conventional rival *El Mundo* generates much more conversation than traffic) generates 34.1% (Fig. 2).

According to media representatives interviewed, the vast majority of Spaniards now prefer to consume news content on the run: more than 70% of the traffic for the sample was mobile. Mobile access peaks at 10 pm and remains high throughout weekends and holiday periods. Morning consumption, on the other hand, tends to take place via desktop computers.

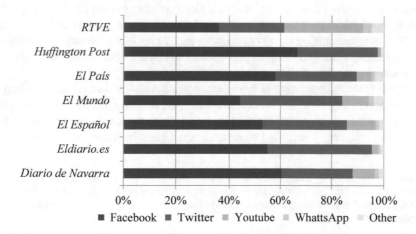

Fig. 2 Traffic from social media. Similarweb

The growing use of mobile devices has clear implications for the future of news organizations. Mobile users are more likely to surf the news than individuals using home computers, consuming an average of only 3.1 pages per session and over half (56.2%) of this audience connects for less than 30 s. In other words, mobile consumption (primarily via smartphones) tends to be much shorter in terms of page views and time frames than desktop consumption. The representative of *The Huffington Post* interviewed for this study observed:

> Mobile-user attention is really brief. We're talking about an average of a minute and thirty seconds per story. It's true there are plenty of short three-paragraph articles a reader can absorb in a flash. But I see this, in any case, as one more sign that what the majority of people do is read a story headline, take a quick look at the related photo and the opening line and move on. (interviewee *The Huffington Post*, pers. comm., April 2019)

News organizations' ability to track and analyze the public's growing habit of consuming content in bits and pieces is undoubtedly having a knock-on effect on the way that news stories are being composed. A representative of the *Diario de Navarra* recounted (Fig. 2):

> We have a tool that allows us to measure not only the overall time visitors spend on our site but also the time they remain fixed on a given scroll position, approximately how long it should take them to read an entire article and the amount of time they really devote to a piece. On the basis of this data we can compare the percentages of shorter and longer articles that actually get read. On the average this works out to around 35 or 40 percent. (interviewee *Diario de Navarra*, pers. comm., February 2019)

4   Shared Newsroom Strategies

Generally speaking, the conventional news enterprises covered by this study were conscious of the impact that the widespread use of mobile devices and social networking sites was having on news consumption habits, had successfully weathered their respective analogue to digital transitions and had sufficient financial and human resources to meet this challenge.

Unlike their digital-native competitors, conventional news outlets have had to adapt longstanding routines to accommodate new online editions. Editorial meetings at which current events are analyzed and contextualized and decisions taken concerning content, format and placement are a case in point. At *El Mundo*, for instance:

> The morning meeting is largely devoted to the digital edition whereas the afternoon session starts off with a discussion about the front page content of the print edition and Web matters and the next day's homepage content are the last things on the agenda. (interviewee *El Mundo* pers. comm., March 2019)

Newspapers like *El País* and *El Mundo* have created job profiles that have helped them to segue from the classic routines of conventional print journalism to new modes of content creation and dissemination. In such scenarios, new professionals work hand in hand with those carrying out more traditional roles, ongoing training programmes keep employee skills up to date, new technologies are exploited, multi-disciplinary teams are formed and new narratives are developed. The representative of *El Mundo* told researchers:

> We hold SEO meetings, a social media staffer comes and gives us a brief rundown of what in particular people have been doing searches on that morning, the topics generating the most traffic on social networking platforms and even the topics generating the most traffic to competitors. We then review our homepage, determine whether the topics being covered there are working, what, if anything, needs to be tweaked and what is performing particularly well and could be worth beefing up before reviewing the traffic being generated by news stories and figuring out which are the hottest topics. (interviewee *El Mundo*, pers. comm., March 2019)

The findings of this study nevertheless indicate that some media outlets in Spain continue to maintain two separate newsrooms, each of which operates according to its rules and customs, a circumstance that obviously hinders the development of even the most basic multiplatform strategy. Some sector professionals cite the differing paces of work in print and digital journalism as a justification for this segregation. The representative of the *Diario de Navarra* interviewed explained:

> Unfortunately, we haven't managed to unify our newsrooms. Each one has its own rhythm. Print newsroom staffers are busiest in the afternoon with their daily deadline and stick to a fairly traditional routine. The work schedule in the digital newsroom is longer, starting at six in the morning and lasting until midnight with the exception of special occasions that require a night shift. (interviewee *Diario de Navarra*, pers. comm., February 2019)

One must therefore conclude that despite a general industry shift towards the fusion of separate and independent newsrooms into single, unified units operating in the same space, under the same editorial leadership and sharing common technology (Salaverría and Negredo 2008), outlets reluctant to take the leap are demonstrating that in certain circumstances maintaining separate print and digital newsrooms can be a viable option (Rondón and Leyva 2017).

The reorganization of newsrooms implies hiring new staffers who know how to (a) multitask (Hamilton 2016), (b) use the latest generation of tools for creating new types of narratives (Paulussen 2016) and (c) take full advantage of the opportunities that social networking platforms offer (Hermida 2016; Jensen 2016).

The media organizations examined for this study felt it was important to recruit individuals with relevant technological training capable of creating quality content in line with industry standards and willing to continually update their skills. The *Diario de Navarra* and *El Español* both consider writing skills important:

> The first thing on our checklist when recruiting a new employee is determining whether a candidate is a good journalist. We know that people are capable of picking up digital skills on the job fairly quickly. (interviewee *Diario de Navarra*, pers. comm., February 2019)

> We seek versatile people who are quick on their feet and have a sound understanding of syntax. A candidate must, above all, be a journalist (...) and have a journalistic mentality. (interviewee *El Español* pers. comm., April 2019)

Some news enterprises need professionals with very specific skills. *RTVE*, for example, is interested in people "well versed in multimedia and editing who are good at visualizing data" and considers "a good command of HTML to be a major plus". Others search the labor market for professionals with non-journalistic profiles. According to the spokesperson for the *Diario de Navarra* researchers interviewed, that newspaper is always looking for "web designers, SEO specialists, technically oriented professionals such as developers and digital marketing specialists".

Several of the individuals interviewed for this study stated that age was also an important factor. The interviewee from the *Diario de Navarra* observed:

> It's a fact that the younger people we're hiring now enter with far broader digital skill set. (...) Staffers who have been here for twenty years were not exposed to computers when they were university students. The only tool they knew how to use when they started out here was a typometer. (interviewee *Diario de Navarra*, pers. comm., February 2019)

Journalists today have no option other than to accept the challenge of mastering and using digital tools: digital technology is essential to their work and broadens the horizons of their profession (García Avilés 2007).

People with a good working knowledge of how to implement SEO and social media strategies appear to be in high demand. The representative of the *Diario de Navarra* who participated in this study explained:

> We consider SEO a high priority area: our Web personnel have solid training and, in fact, we've had a SEO expert in the newsroom for a year now. Once we finished that project we realized that we didn't really need a SEO person to review every word we wrote and that individual journalists needed to internalize the concept. (interviewee *Diario de Navarra*, pers. comm., February 2019)

The majority of the news organizations examined have developed SEO strategies, because as *RTVE* points out, "a media organization doesn't exist if people don't see it". The spokesperson for *El Mundo* recounted:

> We've worked a lot with news writers to ensure that everyone has a basic grasp. In digital, news stories must contain keywords, headlines need to be tagged in HTML, etc. We've worked with every section, explaining, providing hands-on support and teaching people about SEO, and this has made a vital difference. This is obviously easier to do with younger staffers, but everyone has had to gear up. It's been harder to educate older print journalists who didn't understand why you were changing their headlines. Now we're over the hump, but I remember some pretty big confrontations years ago with sports and culture writers who tended to come up with very poetic headlines: I hear you [I'd tell them] but the point is that people aren't reading you. We're losing a big opportunity to attract traffic and what's important here is that the articles you write get read, so you need to change your tack. Start the headline like this followed by two dots and then add your poetic title. This was evidently harder to get across to veterans, but we did it with them too and it worked out well. (interviewee *El Mundo*, pers. comm., March 2019)

Another thing the news outlet examined for this study had in common in an addition to the experience of overcoming their journalists' initial doubts concerning the benefits of SEO is a strong conviction that journalists should actively promote their work via social media. While all believed, like *The Huffington Post*, that "journalists are in better position than anyone else to sell the content they generate", the representative of the *Diario de Navarra* interviewed emphasized that convincing journalists to assume this responsibility requires time and effort:

> As far as social media goes, it depends a little on whether a person is into it or not. At the beginning, we didn't obligate anyone to use or stop using it. What we have done is offered training and developed a few guidelines regarding what people could and couldn't do. Everyone understands it opens up new opportunities to disseminate content. (interviewee *Diario de Navarra*, pers. comm., February 2019).

5   Content Creation and Dissemination

Findings indicate that the fight to capture audience attention minute by minute in a broad range of formats has generated new narratives as well as new practices. The representative of *The Huffington Post* interviewed by the authors explained the dilemmas that newspapers face today when developing content:

> What audiences want to read is insubstantial stuff about what's happening on television, accidents and robberies, things that have gone viral on the Internet, etc. If you publish an article about the eight-year war in Syria, you'll be lucky if as many as a thousand people read it. An article about what went on last night during a TV show like *El Hormiguero*, on the other hand, will get read by 150,000. And the difference lies in the fact that the latter could be churned out in ten minutes and the former about Syria took several days to write. (interviewee *The Huffington Post*, pers. comm., April 2019)

Some news organizations develop sophisticated methods for tracking the performance of content published that contribute to an outlet's success. The spokesperson for *El Mundo* reported:

> Editors and section chiefs review the results on a daily basis to analyze factors beyond quantity, such as the number loyal users who have read a given article, the time site visitors took to read the story and whether or not they read the entire piece. This information guides our judgements about what type of topics have the most potential appeal to readers, which should have more weight, which should be changed, etc. (interviewee *El Mundo*, pers. comm., March 2019).

In other words, the thrust of content creation strategies being pursued by the news organizations analyzed oscillates between a purely quantitative focus on reaching the greatest number of readers possible and a more qualitative focus centered on offering higher quality content targeting mainly loyal readers. At *El Español*, for example:

> Some sections are interested in traffic and others are more preoccupied with building influence, quality and brand image, etc. Sections under pressure to generate traffic are used to pump up the page views that guarantee advertising revenue and profitability and what Pedro J. refers to as influence sections have a greater potential to foster a dialogue with more fickle audience segments, generate viral content and have mass appeal. (interviewee *El Español*, pers. comm., April 2019)

Newsroom personnel involved in content creation are gradually interiorizing the fact that news has become a dynamic rather than a static product that needs to be realigned with what the public is interested in consuming at any given moment. The *El Mundo* spokesperson interviewed for this study acknowledged there had been a change of habits at that daily, noting:

> Analytically minded people in the newsroom are quick to detect opportunities for boosting traffic and where we've made a false step that has resulted in slumping numbers and they are now more accustomed to making decisions on the spot. That's to say, taking action the moment an opportunity strikes and not up to an hour later, which means inserting the right key word right away. (interviewee *El Mundo*, pers. comm., March 2019)

Media outlets have acquired a range of tools, many of which are automated, that allow them to position front page content hierarchically according to the interests of the audiences they serve and opt for the headlines that produce the best results. The spokesperson for *El Mundo* informed the authors:

> We have a tool for testing headlines that allows us to try out three or four headlines for a given article that we use a lot for front-page stories. It allows us to release the same article with different headlines to various segments of our readership and we go with the one that people click on the most (…) Before, we'd publish whatever headline a journalist prepared for an interview he or she had conducted but now we test ten different headlines and in-text quotes and readers are the ones who decide which are the most interesting. (interviewee *El Mundo*, pers. comm., March 2019)

Such practices naturally affect the manner in which headlines are formulated and leave online news outlets facing the dilemma of whether they should or should not employ click bait strategies to boost traffic to their sites. The spokesperson for *The Huffington Post* bemoaned the ramifications of this situation:

> We have real-time measurement tools and it's really frustrating to compare how the same story performs with and without click bait, the latter of which generates 50% more reads. We're talking about the very same story! Faced with the possibility of attracting twice the number of readers, what do I do? Do I go with a journalistic headline? I believe there is a happy medium that doesn't entail going whole-hog with click bait but doesn't involve using journalistic headlines either because on the basis of my personal experience here, what doesn't grab people's attention is of little interest. (interviewee *The Huffington Post*, pers. comm., April 2019)

6 Multiplatform and Transmedia Strategies

In addition to monitoring content in real time, digital news outlets employ a variety of transversal (multiplatform, cross-media and transmedia) strategies to boost audience engagement with their brands, a practice referred to in the industry as cross-media branding. Multiplatform strategies are employed to stimulate user engagement in general by means of the dissemination of content via a wide range of formats and distribution channels (press, radio, television, apps, social media, websites). Transmedia strategies, on the other hand, which focus on interaction with content and by definition require active user participation in the form of sharing, producing, forwarding, etc., are used to stimulate spontaneous user engagement with a mass media brand or its products (news shows, entertainment series and other types of programmes).

The possibility of employing such strategies—be they basic (cross-media) or more complex (transmedia)—hinges upon two factors: a news enterprise's business model and other characteristics (e.g. whether it is a print, audiovisual or digital native enterprise) and the type and format of the content involved (standard news coverage, special news reports, entertainment content, audiovisual content, etc.).

Findings indicate that the degree to which media outlets in the sample have adopted a transversal approach to content varies widely in function of the category they fall into. Press enterprises that publish online newspapers tend to pursue multiplatform strategies centered on mobile apps, an option widely perceived as being particularly suited for fostering audience engagement with newspapers and the content they publish (cross-branding). Apps tend to be viewed as multi-distribution mechanisms in the context of news content and as transmedia tools useful for attracting and engaging audiences at a deeper level in the context of entertainment content. The spokesperson from *El País* noted during his interview, "this approach is particularly evident in the case of audiovisual outlets. In our case, we have developed a really strong application (…) but we don't generate content exclusively for the platform". According to the professional contacted at *RTVE*, the public broadcasting company has adopted a multi-faceted strategy:

> We have an application called Informativos TVE (…) that was intended to replicate what was being aired on television (…) whereas transmedia platforms like Playz are managed by digital teams that produce content for a totally different audience. (interviewee *RTVE*, pers. comm., February 2019)

The media organizations in the study sample employ different strategies for different types of news and journalistic content. Transmedia strategies are reserved for in-depth news programming such as special reports, but multiplatform and cross-media strategies are applied in a more or less de facto manner to daily news programming, undoubtedly due to the shorter shelf-life and less ambitious production and distribution efforts this type of content calls for. The *RTVE* spokesperson who participated in this study explained:

> If the topic being covered is considered to have sufficient weight, our lab develops high-impact strategies entailing different formats for every social media platform (…) a more elaborate Web presence with live videos on Instagram, for example, and different live videos on Facebook. (interviewee *RTVE*, pers. comm., February 2019)

The *Diario de Navarra* focuses heavily on tailoring its strategies and the content it offers to its business model, and, above all, to its format model. The fact that certain types of content are only suitable for presentation via one medium limits this digital newspaper's options for pursuing complex strategies for news, entertainment and promotional content that requires a mix of media formats. Others such as *El Mundo* consider multiplatform strategies to be essential even though employing them supposes approaching the production and distribution of news content from a different angle. The spokesperson from that newspaper interviewed made it clear that his organization had made a substantial investment in this area, pointing out that:

> All of our content is responsive (…) we've assimilated the idea that content should be viewable via any type of device (…) our operations, product and social media teams take care of this issue. (interviewee *El Mundo*, pers. comm., March 2019).

There was a general consensus among the professionals interviewed for this study that video content was especially useful in the context of multiplatform and branding strategies. The professional interviewed from *El Español* thought that producing videos with the potential to go viral was important but stressed that they needed to be "videos that boosted visibility and engagement".

The situation at *RTVE* was interesting in terms of the manner in which internal hierarchies not only structured the work flow involved but also influenced the type of advanced cross-media branding strategies were implemented on a daily basis. Especially worth noting is the fact that all initiatives undertaken at *RTVE* involving several media are now handled by a designated transmedia unit.

Various forms of audience participation (content sharing and resending, input, voting, etc.) have been integrated into *RTVE*'s multiplatform (cross-media) content dissemination and promotional activities. The broadcasting company considers its decision to develop a specific strategy for digital content distributed via various platforms to have been a positive move, but is conscious of the risk of focusing too much attention on the Web and failing to maintain a truly convergent, global strategy that exploits both traditional offline and online channels. The professional contacted at *RTVE* noted:

> There is a consciousness of the need to establish new digital strategies for content dissemination that go beyond adapting offline for digital application (…) This is a new facet of our personality (…) However, for the rest of the corporation (…) the development of purely digital strategies has nothing to do with their part of day-to-day business. (interviewee *RTVE*, pers. comm., February 2019)

The unit is aware that the benefits *RTVE* derives from its transmedia initiatives in terms of audience building and audience engagement are not commensurate with the financial and human resources the broadcasting company devotes to them. All of the sector professionals interviewed for this study admitted that their organizations consider the cross-media and transmedia activity they engage in as an outward sign of their ongoing commitment to innovation and a springboard for exploring new ways of doing things. While transmedia is producing some interesting results in the area of entertainment content, organizations like *RTVE* acknowledge that more work needs to be done in order to apply it effectively in the context of news programming and reaching and engaging new audiences. As far as news presentation goes, second screens continue to be used as 'mirror applications' to replicate content.

Capacity building and the organization and management of workflow are major issues for enterprises like *El Mundo*, *El País* and *El Español* whose business models are focused on the dissemination of news content. The professional from digital-native *El Español* interviewed for this study, for example, admitted, "our capacity to generate news content vastly outpaces our capacity to distribute it through a range of channels" (interviewee *El Español*, pers. comm., April 2019).

7 Conclusions

In addition to attending to the more traditional aspects of journalism, online news outlets have spent much of the last decade developing strategies for satisfying the demands of their increasingly active and mobile respective audiences and adapting to their changing news consumption habits.

The ongoing debate regarding the functions and job profiles of professional journalists has become more intense as the business of journalism has evolved. The sector continues to grapple not only with purely technical considerations but other issues as well such as the search for alternative business models and narratives, newsroom reorganization, the new competences journalists are now being expected to demonstrate and the new roles they are being obliged to assume. A novel environment is remoulding the figure of the reporter into a journalist who narrates in multiple formats and ways (multimedia, transmedia, etc.) (López García, Rodríguez and Pereira 2017). This reporter accepts that certain journalistic practices are inevitably being redefined ant that the job profiles emerging in the sector thus are, and will continue to be, dynamic (López García, Rodríguez and Pereira 2017).

Newsroom professionals have learned to carry out tasks related to SEO and the dissemination of content via social media platforms in tandem with other journalistic responsibilities and have developed tools that facilitate the detailed monitoring of audience behavior in real time. The insight gained through such activities nevertheless comes at a price. Journalism now takes place in a market driven by the preferences of consumers accustomed to receiving a constant, dynamic response to their whims, a circumstance that affects not only the manner in which news stories are written but also their on-screen positioning and hierarchy on a digital newspaper's agenda.

In terms of the introduction of new formats, an increasing number of news organizations see newsroom convergence as the first logical step of their digital transitions and are only venturing into multiplatform content once this has been achieved. The emergence of spaces that permit average people to comment on, interact with and recirculate mass media content has had an impact on sector job profiles, professional routines, short- and medium-term planning processes, audience relationships and the role of audiences.

The model now emerging is marked by a trend toward transversality that constitutes the nucleus of sector innovation.

An ever-stronger focus on content promotion and broader dissemination has had a notable impact on how news enterprises perceive themselves, their business models and how their internal operations are organized.

A close examination indicates that the individual idiosyncrasies, business models and visions of media enterprises greatly determine the degree to which they pursue transversal strategies. Although each of the companies studied here has adopted its own particular approach to transversal media, all have explored the possibilities of social media platforms and apps to some extent, using news and entertainment content to strengthen brand engagement and seeking to enhance the visibility of content they offer by disseminating it via as many channels as possible and leveraging the ability of digital audiences to circulate it further share and virilization.

Acknowledgements The authors would like to thank the representatives of the media enterprises interviewed for this study for their participation and insight.

The topic and focus of the chapter proposed is a natural outflow of two projects carried out by the researchers with funding from the Mini0stry of Economy and Competitiveness, the first of which analyzed the role of active audiences and user-generated content in journalism today (CSO2012-39518-C04-01) and the second of which examined the role active audiences play in setting public agenda (CSO2015-64955-C4-1-R). This work is also part of the scientific production of the Consolidated Research Group (A) 'Gureiker', (IT1112-16), funded by the Basque Government.

References

Casero Ripollés A, López Meri A (2015) Redes sociales, periodismo de datos y democracia monitorizada. In: Campos F, Rúas J (eds) Las redes sociales digitales en el ecosistema mediático. Sociedad Latina de Comunicación Social, La Laguna, pp 96–113

Deuze M (2017) Considering a possible future for digital journalism. Revista Mediterránea de Comunicación 8(1):9–18

Dupagne M, Garrison B (2006) The meaning and influence of convergence. A qualitative case study of newsroom work at the Tampa News Center. J Stud 7(2):237–255

García Avilés JA (2007) Nuevas tecnologías en el periodismo audiovisual. Revista de Ciencias Sociales y Jurídicas de Elche 1(2):59–75

García Orosa B, López García X (2016) Las redes sociales como herramienta de distribución on line de la oferta informativa en los medios de España y Portugal. Mediatika 21(40):125–139

Hamilton JM (2016) Hybrid news practices. In: Witschge T, Anderson CW, Domingo D, Hermida A (eds) The Sage handbook of digital journalism. Sage, London

Hermida A (2016) Social media and news. In: Witschge T, Anderson CW, Domingo D, Hermida A (eds) The Sage handbook of digital journalism. Sage, London

Jensen JL (2016) The social sharing of news: gatekeeping and opinion leadership on Twitter. In: Jensen JL, Mortensen M, Ormen J (eds) News across media: Production, distribution and consumption. Routledge, New York

Killebrew KC (2005) Managing media convergence. Blackwell Publishing, Iowa

Larrondo A, Erdal I, Masip P, Van den Bulck H (2016) Newsroom convergence: a comparative study of European public service broadcasting organisations in Scotland, Spain, Norway and Flemish Belgium. In: Franklin B, Eldridge S (eds) The Routledge companion to digital journalism studies. Routledge, London, pp 556–566

Lewis SC, Westlund O (2016) Mapping the human-machine divide in journalism. In: Witschge T, Anderson CW, Domingo D, Hermida A (eds) The Sage handbook of digital journalism. Sage, London

López García X, Pereira X (2010) Convergencia digital: reconfiguración de los medios de comunicación en España. Universidad de Santiago de Compostela, Santiago de Compostela

López García X, Rodríguez AI, Pereira X (2017) Competencias tecnológicas y nuevos perfiles profesionales: desafíos del periodismo actual. Comunicar 25(53):81–90

Masip P, Guallar J, Peralta M, Ruiz C, Suau J (2015) Audiencias activas y periodismo. ¿Ciudadanos implicados o consumidores motivados? Brazilian Journalism Research 11:240–261

McQuail D (2013) Journalism and society. Sage, Los Angeles

Peña S, Lazkano-Arrillaga I, García D (2016) European newspapers' digital transition: new products and new audiences. Comunicar 46:27–36

Paulussen S (2016) Innovation in the newsroom. In: Witschge T, Anderson CW, Domingo D, Hermida A (eds) The Sage handbook of digital journalism. Sage, London, pp 192–209

Rondón LA, Leyva VH (2017) Redacciones integradas en Cuba: entre el reto y la necesidad. Santiago 144:607–623

Salaverría R (2010) Estructura de la convergencia. In: López García X, Pereira X (eds) Convergencia digital: reconfiguración de los medios de comunicación en España. Universidad de Santiago de Compostela, Santiago de Compostela, pp 27–40

Salaverría R, Negredo S (2008) Periodismo integrado. Convergencia de medios y reorganización de redacciones. Editorial Sol90, Barcelona

Siegert G, Förster K, Chan-Olmsted SM, Ots M (2015) Handbook of media branding. Springer, Switzerland

Tosoni S, Carpentier N, Murru MF (2017) Present scenarios of media production and engagement. Lumière, Bremen

Ufarte MJ, Peralta L, Murcia FJ (2018) Fact checking: un nuevo desafío del periodismo. El profesional de la información 27(4):733–741

Ainara Larrondo Ureta PhD. in Journalism (UPV/EHU) and Senior Lecturer at the Department of Journalism II of the University of the Basque Country. Director of Gureiker group (Basque University System, IT 1112-16) and active member of several research projects on Online Journalism and Media Innovation (Spanish MINECO, FEDER, etc.). Her main educational and research interests are the transformation of journalism in the digital environment, the feminism and the media, and the political communication. Visiting researcher at the CCPR of the University of Glasgow.

Koldobika Meso Ayerdi Ph.D. in Journalism (UPV/EHU) and Senior Lecturer. Head of Department of Journalism II at the University of the Basque Country (Spain). Member of Gureiker research group (UPV/EHU) and main researcher of national and international research projects (Spanish Ministry of Economy and Competitiveness—MINECO and the European Regional Development Fund—FEDER). Visiting researcher at the Federal University of Bahia (Brazil). His main research interests are the transformation of journalism in the digital environment, the audiences and the interactivity.

Simón Peña Fernández Ph.D. in Journalism (UPV/EHU) and Senior Lecturer. Dean of the Faculty of Social Sciences and Communication of the University of the Basque Country (Spain). Codirector of Gureiker research group (UPV/EHU) and researcher of national and international research projects (Spanish Ministry of Economy and Competitiveness—MINECO and the European Regional Development Fund—FEDER, Erasums+, Emakunde). His main research interests are the transformation of journalism in the digital environment, the journalism education, and journalism and cinema.

Social Media Guidelines for Journalists in European Public Service Media

Sabela Direito-Rebollal, María-Cruz Negreira-Rey
and Ana-Isabel Rodríguez-Vázquez

Abstract Strategies and productive routines within the information arena are increasingly oriented towards social networks, as journalists use them to connect with the audience, gather information, disseminate content and build their own brand. Their activity on their personal profiles impacts the credibility of the media they work for, which is why these develop policies for the use of such platforms. In this chapter we analyze the social media guidelines developed by the public service broadcasting corporations of the European Union—*BBC, ORF, RTÉ, Sveriges Radio, TVR, YLE, EITB, NDR* and *VRT*—with the objective of clarifying the recommendations for their journalists. The analysis of these documents reveals that organizations place special emphasis on regulating the personal and professional use of social media accounts, the relationship with the audience, reporting, as well as on transparency and the treatment of confidential information.

Keywords Social media · Guidelines · Public service media · Journalistic practices

1 Introduction

Since the beginning of the 21st century, the news industry has faced the challenge of adapting to the radical and disruptive changes caused by the digital shift (Paulussen 2016). The Internet and its social and interactive features have profoundly transformed both newsroom culture and journalistic practices (Zeller and Hermida 2015). In this context, newsrooms are immersed in a "makeover process" (García-Avilés et al. 2014: 10) adapting their strategies and productive routines to

S. Direito-Rebollal (✉) · M.-C. Negreira-Rey · A.-I. Rodríguez-Vázquez
Universidade de Santiago de Compostela, Santiago de Compostela, Spain
e-mail: sabela.direito@usc.es

M.-C. Negreira-Rey
e-mail: cruz.negreira@usc.es

A.-I. Rodríguez-Vázquez
e-mail: anaisabel.rodriguez.vazquez@usc.es

© Springer Nature Switzerland AG 2020
J. Vázquez-Herrero et al. (eds.), *Journalistic Metamorphosis*,
Studies in Big Data 70, https://doi.org/10.1007/978-3-030-36315-4_10

an information environment increasingly oriented towards social networks (Welbers and Opgenhaffen 2019).

The growing popularity of platforms such as Twitter and Facebook and their use in collecting and distributing pieces of news offers new opportunities, but also new challenges for professional journalism (Nielsen et al. 2016; Welbers and Opgenhaffen 2019). As Kramp and Loosen (2017: 213) point out, "journalistic roles were confronted with differentiated tasks and practices" that have influenced not only the work of journalists, but also the responsibilities they have to assume (Adornato and Lysak 2017). As well as their traditional work, editors are now expected to perform new tasks in order to meet the demands of an active and participatory audience that seeks to interact in a reciprocal and bidirectional way with the media themselves, as well as with their staff (Canter 2013).

Given this scenario, individual journalists have started using social media to connect with their audiences, but also to gather information, find ideas and sources, report news, promote stories and build their personal brands (Al-Rawi 2017; Barberá et al. 2017; Duffy and Knight 2019; Paulussen and Harder 2014; Thurman 2018). As their activity on these platforms determines not only their personal image, but also the corporate brand of the company, publishers and managers can put pressure on their journalists to become active users of social networks (Hedman and Djerf-Pierre 2013). However, with an awareness of the risks that improper use of this type of platforms can cause on their reputation and that of their journalists (Lee 2015), some media have started to regulate their journalists' personal use of social networks through policies and guidelines. In this way, they seek to preserve the transparency, the objectivity, the authenticity of the sources, the truthfulness of the contents that are published and shared, and the independence of the media themselves and the people who work for them.

In this chapter we analyze the social media guidelines developed by the European broadcasting corporations. Our main objective is to elucidate the guidelines or recommendations that these corporations make to their journalists in order to guarantee the right usage of their personal social media profiles. Based on the existing literature, five major categories on which media organizations place special emphasis have been identified: regulating the characteristics and activity of the personal and professional accounts of their journalists (personal and professional activity), the appropriate use of content—especially content generated by users—and sources (content and sources), the relationship with the audience and the management of complaints or criticisms (audience relationship management), social media reporting (reporting), as well as transparency in their activities and the treatment of confidential information (transparency). This analysis has focused on nine publicly owned corporations whose social media guidelines are accessible through their websites: the state-owned *BBC* (United Kingdom), *ORF* (Austria), *RTÉ* (Ireland), *Sveriges Radio* (Switzerland), *TVR* (Romania) and *YLE* (Finland), and the regional *EITB* (Spain), *NDR* (Germany) and *VRT* (Belgium).

The chapter begins by examining the pros and cons of the use of social channels in the journalistic field. It then goes on to review those works that have analyzed, from various perspectives, the implementation of social media guidelines within media outlets and news agencies. Through the analysis of the policies designed by the European broadcasting corporations, we were able to identify a series of preventative recommendations that are designed to keep their journalists from damaging, through their personal activity in social networks, core values for public service media such as independence, excellence, responsibility and transparency. Thus, the present study contributes to an emerging field of work that, according to various authors (Adornato and Lysak 2017; Lee 2016), has not yet received the necessary attention from the academy.

2 Pros and Cons of Social Media Usage by Professional Journalists

The rise of social media and its gradual incorporation into newsrooms precipitated the transition towards a new journalism model where content is produced, distributed and consumed through multiple platforms (Kramp and Loosen 2017). Networks such as Facebook, YouTube, Twitter and Instagram and messaging services such as WhatsApp have already become fixtures for many users, especially the youth, to access the news (Newman et al. 2019). According to data from the latest *Digital News Report*, 57% of the population between 18 and 24 years old (Generation Z) and 43% of those between 25 and 34 years old (Generation Y) receive the news through these kinds of platforms (Newman et al. 2019). Following the media social profiles and those of the journalists working on them has also become a way for some users to stay informed (Hermida et al. 2012).

Therefore, social networks have been consolidated as a channel of direct communication with the audience, transforming the classic top-down relationship into an open conversation (Paulussen and Ugille 2008). The establishment of this dialogue with the readers "offers a chance for journalists to achieve greater accountability and transparency" (Lasorsa et al. 2012: 21) and, at the same time, makes the behavior and preferences of news consumers more quantifiable and transparent (Assmann and Diakopoulos 2017).

As well as establishing a conversation with multiple users in the same place—one of the greatest benefits that these types of platforms bring about for journalists, according to Safori (2019)—social media provide the opportunity for journalists to reach new sources beyond the institutional ones (Diakopoulos et al. 2012). Social media thus acquire an outstanding role in the search for sources (Thurman 2018) and information (Paulussen and Harder 2014), and are an essential element for news gathering and reporting, especially around breaking news events (Lee 2016). In fact, as Canter (2013: 474) points out, "major events, whether human or natural, are now most likely to be revealed via social media first, from public on the ground in the heart of the action, before the professional journalists arrive". Citizens become amateur reporters (Noor 2017) who, through their mobile devices and social networks, create content and spread news (Barnard 2018; Bowman and Willis 2003; Campbell

2015). These contents generated by the audience become a fundamental piece "not only for breaking news during a crisis, but also for crowdsourcing information for nonbreaking news stories" (Lee 2016: 109).

The shift in the role and relationship that journalists establish with their readers through social networks helps them improve engagement with their users; while at the same time building their own personal brand (Duffy and Knight 2019; Holton and Molyneux 2017; Knight and Cook 2013). However, this opportunity for journalists to create and share their identity online can also pose a major threat to their careers. The thin line that separates their identity as professionals and as individuals in social networks (Holton and Molyneux 2017) can mean that their activity on these platforms is sometimes disengaged from conventional journalistic values and norms (Domingo and Heinonen 2008; Hedman and Djerf-Pierre 2013). In the words of Lee (2015: 4), "journalists are subject to the influence of social media norms such as personality disclosure and interaction", so that they express opinions, share private life moments, talk about their daily work or speak with other users through their social media profiles (Lasorsa et al. 2012). These types of practices, although they offer a more transparent image of journalists in the personal and work environment (Molyneux and Holton 2015), negatively affect their professional reputation to the extent that the audience perceives these behaviors as a violation of the core journalistic principles of objectivity, impartiality and detachment (Hermida 2013; Lee 2015).

The lack of privacy and the circulation of misleading information are also problems journalists need to consider when using social networks (Safori 2019). An error in a post, a link to a website that contains non-verified information, or the re-distribution of misleading news (Lee 2016) can also compromise the public image of journalists by undermining the intrinsic values of the profession such as verification, accuracy and validity of content and sources (Brandtzaeg et al. 2016).

3 The Integration of Social Media Guidelines into Newsrooms

Having considered the opportunities, and even more so the dangers, that the use of social media by journalists can entail, the media have decided to guide the conduct of their workers on such platforms in order to protect their reputation and credibility as informational brands (Bloom et al. 2016). In some organizations, professionals specialized in the management of corporate social profiles are the ones who advise the rest of the workforce on their use to guarantee an optimal relationship with the public, as well as to verify the sources and content obtained through these platforms (Sacco and Bossio 2017). Such guidance can be complemented with coaching and training for journalists (Bloom et al. 2016; Sacco and Bossio 2017).

However, other media outlets understand the presence and activity that both the organization and their workers make of social media as a matter of governance and therefore choose to control their conduct through policies and guidelines (Sacco and Bossio 2017). Although the contents of these standards and practices vary across media, they all seek to maintain a consistent image and conduct between the corporate profiles and those of their journalists (Safori 2019), to identify the dangers of social media, to avoid any inappropriate content and to ensure that the information cannot be used to challenge the integrity of their reporters, photographers and editors (Podger 2009). In general terms, their approach is more oriented towards prevention than promotion and they support a critical understanding of social media (Lee 2016).

Social media guidelines are documents of varying rigidity that govern the presence of the media and their journalists on those platforms (Bloom et al. 2016) and should be subject to changes and updates (Podger 2009). Sacco and Bossio (2017: 185) define them as "the formalization of organizational jurisdiction over the conduct of journalists and other editorial staff in the use of social media for the news production, dissemination and promotion".

Previous studies on social media usage guides from various media and international news agencies (Adornato and Lysak 2017; Bloom et al. 2016; Lee 2016; Opgenhaffen and Scheerlinck 2014; Safori 2019) revealed some common ground that media organizations seek to maintain control over. According to the authors, these guidelines limit the personal and professional use that journalists make of their profiles and define boundaries to safeguard their credibility and impartiality—user identification and relationship with the company, privacy, personal opinion, political ties, etc. They also draw lines on the integration of social networks in the production of information contents, especially in terms of involvement and interaction with the public, and on the verification of sources and user-generated content. Moreover, they provide guidelines on transparency, confidentiality regarding internal media information, response to criticisms and reputational crises, the management and publication of latest news and exclusives, as well as on the accuracy of contents and rectifications.

There are currently few media that have specific codes of conduct for journalists on social media, and still fewer have a written policy (Adornato and Lysak 2017; Opgenhaffen and Scheerlinck 2014; Sánchez Gonzales and Méndez Muros 2015), in which case publishers communicate the guidelines verbally or extrapolate the deontological principles of their general code (Bloom et al. 2016; Opgenhaffen and Scheerlinck 2014). By 2012, these guides were becoming a feature in many newsrooms (Adornato and Lysak 2017). Among the media organizations that use them are newspapers and leading cybermedia—*The New York Times, The Washington Post, The Guardian, The Globe and Mail, El País* or *El Comercio*—public and private broadcasters—*BBC, NPR, ABC, Channel 4, VRT* or *RTÉ*—as well as news agencies—*Reuters, EFE, Associated Press* or *Agence France Press*.

At the same time that media organizations were establishing action rules on social networks that were based on their editorial criteria and values, the social media platforms themselves were preparing manuals for journalistic use. In 2012, Twitter published an entry on their blog with examples of useful practices for the collection and dissemination of information (Luckie 2012), whilst Facebook launched a guide a year later (Lavrusik 2013), advising journalists on the creation of their personal pages, on breaking news coverage, reporting and contact with sources, and on the creation of a participatory community. Today, these modest standards have been transformed into complex projects such as the *Facebook Journalism Project*, which integrates the media themselves for the development of new technological solutions, also providing training and resources for editors and journalists, as Twitter Media also offers.

4 Principles of the Social Media Guides of Public European Broadcasting Media

The guidelines for the use of social networks of public European broadcasting media that were analyzed (BBC 2015; EITB 2014; NDR 2012; ORF 2012; RTÉ 2013; Sveriges Radio 2013; TVR 2012; VRT n.d.; YLE 2015) present variations in terms of the extension and the specifics of the standards, the social platforms included and the distinction between personal and professional profiles. Nonetheless, it is possible to organize these guides using a series of categories implemented in previous studies (Adornato and Lysak 2017; Safori 2019). These include the confrontation of personal and professional use of social profiles, relating to sources and content, relationship management and audience critics, reporting and transparency.

4.1 Personal and Professional Activity

All social media guidelines that were studied refer in their regulations to the limits that professionals of the respective corporations must abide by in their personal accounts.

One of the main policies that all media outlets insist on is that personal profiles remain consistent with corporate principles and values. Similarly, they demand that journalists maintain political impartiality and do not position themselves in controversial or polemical issues on the information agenda. Despite preserving their journalists' freedom of expression, the media warn of the potential damage that this could cause both in their credibility and in their corporate profiles.

Regarding the expression of personal opinions, the journalists of the *TVR* are obliged to include the following disclaimer: "any opinions expressed here are personal and do not represent Romanian television". The *NDR* also recommends clarifying that the comments published are personal, and enforces restrictions to the creation and dissemination of content in personal profiles using the company's computers. The *VRT* recommends being mindful of impartiality principles, which apply not only to the publications, but also to the groups the journalists belong to, the likes they receive, the accounts they follow, etc.

The impact that journalists' personal publications can have on the image of the media group that they work for gives rise to different strategies in their identification in social networks. Thus, while the *BBC* does not allow reference to their working relationship on the profile description, the *Sveriges Radio* obliges journalists to make their biography clear, recognizable and related to the organization.

4.2   Sources and Content

The broadcasting corporations studied encourage their professionals to make use of social networks in their journalistic work when they search for information, identify witnesses or people involved in a particular story, or to find images produced by users. But the treatment of sources and content on social media needs to be addressed under a series of journalistic principles of conduct that the media refer to in their guides.

Therefore, the rules of the *EITB, Sveriges Radio, ORF, RTÉ* and *BBC* focus on the need to verify both the content and the information generated in social media around a news event, as well as the accounts that are behind such publications. As the media deontological principles are extrapolated to social networks, the guides also mention the need to contrast sources and protect them.

The broadcasting organizations also remind journalists that they should be cautious with their use of user-generated content. In addition to their verification, the guidelines stipulate that such content cannot be published without the author's consent and that copyright must be respected. In this sense, the rules of the *YLE* establish that the ethical norms applied to other professional activities within the corporation are extrapolated to social media and that the editorial responsibility remains the same.

Regarding the relationship that the professionals maintain through social networks with users or sources, the *NDR* recommends keeping distance and impartiality. While favoring the participation and collaboration of users, the *BBC* recommends that the audience is not encouraged to take risks on behalf of the corporation.

4.3   Audience Relationship Management

Together with the search for sources and information, the ability to converse in real time with the audience is one of the greatest advantages that social networks bring to journalists. Therefore, it is not surprising that public service media encourage journalists to actively engage with their followers, not only through the organization's official accounts, but also through their own personal profiles.

In fact, the guides of the *ORF, YLE* and *Sveriges Radio* all state the need for their journalists to be part of social networks in order to connect with their users and respond to their questions or comments. Moreover, this dialogue with the audience can also become an opportunity to learn more about their likes and interests—as specified by the Austrian corporation *ORF*—or even a way to disseminate the content of the organization—as reflected in the guide of the *NDR*. Another point that is especially emphasized within the policies of the *Sveriges Radio* and *ORF* is the idea of journalists as moderators who are able to control the nature of social conversations.

The *ORF* also underlines the need for their workers to maintain impartiality, objectivity and independence in their feedback with the audience, particularly stressing the importance of avoiding expressions of political preferences or opinions.

In addition to the ideas and corrections that users may contribute, the criticisms they can make either to the organization or the work of their journalists are included as a highly relevant section in most of the guides analyzed. In this regard, the rules of *Sveriges Radio, NDR, ORF, EITB* and *VRT* emphasize that it is necessary to accept and address criticism, avoiding controversies or discussions with the audience. In the event of a disagreement, the *NDR* recommends their journalists be responsible and transparent, while the *Sveriges Radio* points out the importance of acting fast—but never from a panicked perspective—and being personal, clear and truthful. Being polite rather than reacting aggressively, for example by blocking a user, is the general guideline established by the *BBC*. The *EITB* focuses on the need to calculate the impact and scope of the criticism, as well as the type of person behind it.

4.4 Reporting

Regarding the role of journalists as reporters on social networks, the *EITB, NDR* and *RTÉ* state in their guides that the content produced by the media belongs to corporations and that it should be published in corporate profiles rather than personal ones. Similarly, they remind their staff that they cannot decide on the management of the information they produce—*RTÉ* warns that they reserve the right to ask their employees to delete certain posts from their personal profiles—recommend that publications be shared from corporate social accounts and that the hashtags used are the ones determined by the organization, and also that the quality of the content is carefully considered. Regarding the style of the publications, the *TVR* urges professionals to respect their visual identity guide in their personal accounts.

In relation to breaking news, both the *BBC* and *VRT* prioritize verification and rigor over speed, and make it clear that journalists must not broadcast news on their personal accounts without previously notifying the organization and adhering to their decisions on how to manage the publication.

4.5 Transparency

Clarity and transparency of information are core values within public service media. This is reflected in the social media guides of the corporations analyzed here. However, although they mention the need for their staff to be transparent in their relationships with their audiences, and in how they share content and sources, they also place special emphasis on the treatment of the organization's confidential information.

The policies of all public broadcasters include a standard that explicitly prohibits the disclosure of such information. This implies not revealing the organization's strategy, their internal policies—as specified by the *TVR, NDR* and *YLE*—financial information or details about their suppliers and partners—as the *NDR* maintains—and confidential sources—a guideline that is shared by both the *NDR* and *RTÉ*.

Likewise, although public broadcasters respect the freedom of expression of their journalists, they recommend that social networks not be used as a space to criticize their peers or professionals who work for other media. With this in mind, the *ORF* adds that negative and unsubstantiated comments or opinions on the organization or the content they offer must not be issued. *RTÉ*'s policy is that their journalists must not publish details about their private lives on their personal accounts. The *TVR* shares this recommendation and urges that the privacy of their colleagues be respected.

5 Conclusions

The impact of social networks on the practices and routines of journalists has been the focus of much attention coming from professionals and academics in recent years. Their use as a platform for access, collection and distribution of news has transformed social networks into an effective tool in the daily activity of reporters, as well as a direct route for interaction and connection with users. Meanwhile, media organizations see the presence of their journalists on social networks as an opportunity to expand their reach and visibility, increasing the number of readers, driving traffic to their website and strengthening brand loyalty (Lee 2015; Opgenhaffen and Scheerlinck 2014; Sacco and Bossio 2017).

However, the use of social media can be seen as a 'double-edged sword' (Lee 2015). The interpersonal and subjective logic of platforms such as Twitter (Duffy and Knight 2019; Welbers and Opgenhaffen 2019) can cause journalists to deviate from conventional journalistic values and norms when they perform certain activities through their social media profiles. Given the potential risks that this entails, European public service broadcasting corporations have joined together in order to produce guides for the use of social networks and to steer their staff on how to use their personal accounts.

As has already been observed in previous studies on media and international agencies policies, the analysis of documents published by the *BBC, ORF, RTÉ, Sveriges Radio, TVR, YLE, EITB, NDR* and *VRT* reveals the media's desire to protect the image and credibility of their informational brand, as well as to enforce compliance with the code of ethics on social platforms.

The promotion of political and ideological independence is one of the central values of public service media. Therefore, in their social media policies, they warn that journalists' expressing their personal opinions, and the relationships they establish through social networks with other users, can compromise their impartiality, as well as their adherence to editorial standards. In order to maintain their reputation and informative credibility, all corporations recommend their professionals not to position themselves in political or social matters and act in accordance with the principles and values of the media.

Verifying the information and sources as well as the authenticity of the contents generated by users, managing the permits and publication rights, aligning with the organization before the publication of the latest news and the dissemination of content produced by the journalist, and responding appropriately to comments and complaints are all standards included in most of the guides analyzed.

Beyond aspects linked to journalistic practices, the media address corporate issues, such as the protection of internal and confidential information, the respect for the organization and their team, public or non-public association with the corporation or restriction for publication and use of personal profiles from the actual company.

Although the documents analyzed differ in their length, structure, frequency of update and in specific details, they all reflect an interest in the integration of their professionals within social networks. Nonetheless, the nature of the recommendations and how these are expressed are mainly aimed at forms of journalistic misconduct, however corporations choose to define it. Notwithstanding the guidelines explicitly state that their staff's freedom of expression is respected, the publication of guides is nothing but a tool to control all the activity related to corporations in social media.

The review of the content and formats of social media policies for journalists in the European public broadcasters opens new avenues for future research, such as the ways in which these policies might come into conflict with the real actions of journalists, their implementation in the newsrooms and the extent to which they actually meet the needs of the teams.

Acknowledgements This article has been developed within the research project Digital native media in Spain: storytelling formats and mobile strategy (RTI2018-093346-B-C33) funded by the Ministry of Science, Innovation and Universities (Government of Spain), Agencia Estatal de Investigación, and co-financed by the European Regional Development Fund (ERDF), and it is part of the activities promoted by Novos Medios research group (ED431B 2017/48), supported by Xunta de Galicia. The author Sabela Direito-Rebollal is a beneficiary of the Faculty Training Program funded by the Ministry of Science, Universities and Innovation (FPU15/02557).

References

Adornato A, Lysak S (2017) You can't post that!: social media policies in U.S. television newsrooms. Electron News 11(2):80–99

Al-Rawi A (2017) News values on social media: news organizations' Facebook use. Journalism 18(7):871–889

Assmann K, Diakopoulos N (2017) Negotiating change: audience engagement editors as newsroom intermediaries. ISOJ J 7(1):25–44

Barberá P, Vaccari C, Valeriani A (2017) Social media, personalisation of reporting, and media systems' polarization in Europe. In: Barisione M, Michailidou A (eds) Social media and European politics: rethinking power and legitimacy in the digital era. Palgrave Macmillan, London, pp 25–52

Barnard SR (2018) The pros and cons of pro-am journalism: breaking news during the #Boston-Marathon bombing and beyond. In: Barnard SR (ed) Citizens at the gates. Palgrave Macmillan, Cham, pp 83–100

Bloom T, Cleary J, North M (2016) Traversing the "Twittersphere" social media policies in international news operations. J Pract 10(3):343–357

Bowman S, Willis C (2003) We media: how audiences are shaping the future of news and information. The Media Center at the American Press Institute. Retrieved from http://www.hypergene.net/wemedia/download/we_media.pdf. Accessed 11 June 2019

Brandtzaeg PB, Lüders M, Spangenberg J, Rath-Wiggins L, Følstad A (2016) Emerging journalistic verification practices concerning social media. J Pract 10(3):323–342

BBC (2015) Social media guidance for staff. Retrieved from http://news.bbc.co.uk/2/shared/bsp/hi/pdfs/26_03_15_bbc_news_group_social_media_guidance.pdf. Accessed 11 June 2019

Campbell V (2015) Theorizing citizenship in citizen journalism. Dig J 3(5):704–719

Canter L (2013) The interactive spectrum: the use of social media in UK regional newspapers. Convergence 19(4):472–495

Diakopoulos N, De Choudhury M, Naaman M (2012) Finding and assessing social media information sources in the context of journalism. In: Proceedings of the SIGCHI conference on human factors in computing systems. University of Texas, Austin, 5–10 May 2012

Domingo D, Heinonen A (2008) Weblogs and journalism: a typology to explore the blurring boundaries. Nordicom Rev 29(1):3–15

Duffy A, Knight M (2019) Don't be stupid: the role of social media policies in journalistic boundary-setting. J Stud 20(7):932–951

EITB (2014) EiTB en las redes sociales. Retrieved from https://images14.eitb.eus/multimedia/corporativo/documentos/EiTB-en-las-redes-socales.pdf?_ga=2.202067352.808325400.1559127054-1870109587.1559127054. Accessed 11 June 2019

García-Avilés JA, Kaltenbrunner A, Meier K (2014) Media convergence revisited: lessons learned on newsroom integration in Austria, Germany and Spain. J Pract 8(5):573–584

Hedman U, Djerf-Pierre M (2013) The social journalist: embracing the social media life or creating a new digital divide? Dig J 1(3):368–385

Hermida A (2013) #Journalism: reconfiguring journalism research about Twitter, one tweet at a time. Dig J 1(3):295–313

Hermida A, Fletcher F, Korell D, Logan D (2012) Share, like, recommend: decoding the social media news consumer. J Stud 13(5–6):815–824

Holton AE, Molyneux L (2017) Identity lost? The personal impact of brand journalism. Journalism 18(2):195–210

Knight M, Cook C (2013) Social Media for journalists: principles and practice. Sage, London

Kramp L, Loosen W (2017) The Transformation of journalism: from changing newsroom cultures to a new communicative orientation? In: Hepp A, Breiter A, Hasebrink U (eds) Communicative figurations: transforming communications in times of deep mediatization. Palgrave Macmillan, Cham, pp 205–239

Lasorsa DL, Lewis SC, Holton AE (2012) Normalizing Twitter: journalism practice in an emerging communication space. J Stud 13(1):19–36

Lavrusik V (2013) Best practices for journalists on Facebook. In: Journalists on facebook. Facebook. Retrieved from https://www.facebook.com/notes/journalists-on-facebook/best-practices-for-journalists-on-facebook/593586440653374/?__tn__=HH-R

Lee J (2015) The Double-edged sword: the effects of journalists' social media activities on audience perceptions of journalists and their news products. J Comput Med Commun 20(3):312–329

Lee J (2016) Opportunity or risk? How news organizations frame social media in their guidelines for journalists. Commun Rev 19(2):106–127

Luckie M (2012) Best practices for journalists. In: Blog Twitter. Retrieved from https://blog.twitter.com/en_us/a/2012/best-practices-for-journalists.html

Molyneux L, Holton A (2015) Branding (health) journalism. Dig vJ 3(2):225–242

NDR (2012) NDR social media guidelines. Retrieved from https://www.ndr.de/der_ndr/daten_und_fakten/handbuchorganisation138.pdf. Accessed 11 Aug 2019

Newman N, Fletcher R, Kalogeropoulos A, Nielsen RK (2019) Reuters Institute Digital News Report 2019. Retrieved from https://reutersinstitute.politics.ox.ac.uk/sites/default/files/2019-06/DNR_2019_FINAL_1.pdf. Accessed 21 May 2019

Nielsen RK, Cornia A, Kalogeropoulos A (2016) Challenges and opportunities for news media and journalism in an increasingly digital, mobile, and social media environment. Retrieved from https://rm.coe.int/16806c0385. Accessed 15 May 2019

Noor R (2017) Citizen journalism versus mainstream journalism: a study on challenges posed by amateurs. Athens J Mass Media Commun 3(1):55–76

Opgenhaffen M, Scheerlinck H (2014) Social media guidelines for journalists: an investigation into the sense and nonsense among Flemish journalists. J Pract 8(6):726–741

ORF (2012) Social-media-guidelines für ORF-JournalistInnen. Retrieved from https://zukunft.orf.at/rte/upload/texte/2012/social_media_guidelines_orf_final.pdf

Paulussen S (2016) Innovation in the Newsroom. In: Witschge T, Anderson CW, Domingo D, Hermida A (eds) The SAGE handbook of digital journalism. Sage, London, pp 192–206

Paulussen S, Harder R (2014) Social media references in newspapers: Facebook, Twitter and YouTube as sources in newspaper journalism. J Pract 8(5):542–551

Paulussen S, Ugille P (2008) User Generated content in the newsroom: professional and organisational constraints on participatory journalism. Westminster Pap Commun Cult 5(2):24–41

Podger PJ (2009) The limits of control: with journalists and their employers increasingly active on social media sites like Facebook and Twitter, news organizations are struggling to respond to a host of new ethics challenges. Am J Rev 31(4):32–38

RTÉ (2013) RTÉ social media guidelines. Retrieved from https://static.rasset.ie/documents/about/social-media-guidelines-2013.pdf. Accessed 11 June 2019

Sacco V, Bossio D (2017) Don't Tweet This! How journalists and media organizations negotiate tensions emerging from the implementation of social media policy in newsrooms. Dig J 5(2):177–193

Safori AO (2019) Journalist use of social media: guidelines for media organizations. J Soc Sci Res 5(4):1061–1068

Sánchez Gonzales H, Méndez Muros S (2015) Las guías de uso de medios sociales: regularización periodística y calidad informativa. Estudios sobre el mensaje periodístico 21:143–154

Sveriges Radio (2013) Social Media. A handbook for journalists. Retrieved from https://www.ebu.ch/files/live/sites/ebu/files/Publications/Swedish_Radio-Social_Media-Handbook-for-Journalists.pdf. Accessed 11 June 2019

TVR (2012) Reglementari privind blogurile si toate celelalte forme de webcontent personal pentru angajatii SRTv. Retrieved from http://www.tvr.ro/reglementari-privind-blogurile-si-toate-celelalte-forme-de-webcontent-personal-pentru-angajatii-srtv_6.html#view. Accessed 11 June 2019

Thurman N (2018) Social media, surveillance, and news work: on the apps promising journalists a "crystal ball". Dig J 6(1):76–97

VRT (n.d.) Tien geboden voor sociale media. Retrieved from https://www.vrt.be/nl/over-de-vrt/beleid/beroepsethiek/10-geboden-voor-sociale-media/. Accessed 11 June 2019

Welbers K, Opgenhaffen M (2019) Presenting news on social media. Dig J 7(1):45–62

YLE (2015) Sosiaalisen median toimintalinjaukset. Retrieved from https://yle.fi/aihe/artikkeli/sosiaalisen-median-toimintalinjaukset. Accessed 11 June 2019

Zeller F, Hermida A (2015) When tradition meets immediacy and interaction. The integration of social media in journalists' everyday practices. Sur le journalisme 4(1). Retrieved from https://surlejournalisme.com/rev/index.php/slj/article/view/202

Sabela Direito-Rebollal Ph.D. Candidate at Universidade de Santiago de Compostela (USC). She holds a Degree in Audiovisual Communication (USC), a Master's Degree in Communication and Creative Industries (USC) and a Degree in Movie and TV Script from the Madrid Film Institute. She was visiting scholar at the University of Hull (United Kingdom) and the Vrije Universiteit Brussel (Belgium). Her research focuses on innovation, audience trends and programming strategies of European public service media.

María-Cruz Negreira-Rey Ph.D. Student in Communication and Contemporary Information at Universidade de Santiago de Compostela and Member of Novos Medios research group (USC). She develops her research in the areas of online journalism and media of proximity, focusing on the hyperlocal online media in Spain and Portugal.

Ana-Isabel Rodríguez-Vázquez Ph.D. in Communication at Universidade de Santiago de Compostela. She has a BA in Information Science from Universidad Complutense de Madrid (UCM). After more than a decade working in the media, she started to work at the USC in 2003. She belongs to the Audiovisual Studies research group (GEA) at the USC, and her main fields of research have to do with information, programming and audiences.

Reinventing Television News: Innovative Formats in a Social Media Environment

Jose A. García-Avilés

Abstract Television news consumption is declining, while videos on YouTube and social media are on the rise. Newscasts seem unable to attract a young audience that actively engages online. The purpose of our study is to establish how television news could be renovated by selecting the experts' proposals and by exploring case studies of innovative audio-visual news formats. We posed two research questions: (a) How television newscasts could be renewed in the stages of production, edition, presentation and distribution? (b) Which news formats are innovating in storytelling and connecting with audiences? Our findings suggest that young audiences must be taken seriously by broadcasters for their long-term future. The experts suggest ways to experiment with new languages, diversify the news offer, foster journalists' communicative skills and increase interaction with users.

Keywords Newscasts · News formats · Innovation · Journalism · Television

1 Newscasts Are Becoming Increasingly Irrelevant

Television is losing its dominant position as the main news medium. Traditional broadcast news consumption is declining, while online video in YouTube, websites and social media is on the rise. In the United States, television consumption has declined by 3% since 2012, while the audience of newscasts is getting increasingly older: the average age of *CNN* viewers is 61 years, of *MSNBC* is 63, of *CBS* and *ABC*, 64, and of *Fox News* is 68 years, according to Nielsen (2018).

In Spain, the consumption of linear television reaches 236 min a day. It continues to be the main source of information and leisure for most of the population, although the growth of non-linear television is unstoppable. Broadcasts news still reaches relevant audiences, mostly in news, sports and live shows. Considering the top five generalist national channels, 16 million viewers watch regularly a newscast in Spain and there are several newscasts among the most watched daily programs (Barlovento Comunicación 2019).

J. A. García-Avilés (✉)
Miguel Hernández University, Elche, Spain
e-mail: jose.garciaa@umh.es

J. Vázquez-Herrero et al. (eds.), *Journalistic Metamorphosis*,
Studies in Big Data 70, https://doi.org/10.1007/978-3-030-36315-4_11

However, the decline in television consumption among young people is more pronounced for news programs. Although many people over 55 consume television regularly, the news are no longer of interest to most of the under 30 years old, who have practically abandoned newscasts (Newman et al. 2018). Some studies show that youngsters are not interested in broadcast news and that it is difficult to reconcile these age segments with newscasts, because the format requires minimal attention that these young multitaskers are not willing to give (Drok et al. 2018).

Since its inception, television has incorporated technological advances in news gathering, production and presenting. But the essence is always the same: a presenter who conducts a narrative of fragmentary pieces that make up a supra-story of current events that intent to be decisive for public conversation. In the horizontal communication ecosystem of the Network Society, the newscast's narrative has several problems (Canter 2018). It is a vertical communication, which originates from a professional authority, the anchor, who speaks directly to the audience. A newscast cannot be presented by a social media 'friend'. It provides an edited version of the news, both structured and hierarchical, just the opposite of what interactive users are supposed to want. It represents immediacy through live broadcasting, but it cannot compete with the instantaneity of the Internet (Kalogeropoulos and Nielsen 2018). It is a one-way narration that does not allow interaction with the audience and it has lost the monopoly of today's pictures, which originate in social media. For all these reasons, newscasts may be unattractive to the youngest.

As Terán (2017) argues, daily newscasts currently are a summary of press conferences and soundbites, a compilation of pieces cut by the same pattern and homogeneous contents so that the news provided by the different channels resemble each other a lot. Thus, newscasts are an obsolete format in the current environment of instant, on-demand and cross-platform information (Noguera Vivo 2018).

If the television networks do not react to the decrease in the consumption of broadcast news, newscasts will soon be irrelevant. The formula that for decades has supported the newscast based on a summary of the highlights of today in the midday and evening editions has become obsolete, because in the age of social media and a continuous 24-h news flow, people already know the news. Therefore, newscasts need to reinvent themselves if they do not want to become irrelevant.

2 Innovating in Television News Formats

Innovation always involves risks. It brings about change or discontinuity, both in terms of the transformation of an idea into a reality and in terms of its impact on the organizations and society. To innovate implies developing a new concept, product or service in a specific market, in a disruptive way, that is, that this novelty alters the traditional way in which things have been done (García-Avilés et al. 2018a). Media innovation does not only involve the development of a technological process or service, but the effect of improving a news product, increasing the company's reputation or providing a competitive advantage (García-Avilés et al. 2018b).

Broadcast news innovation requires a multidisciplinary approach to connect with the needs of the public, because audiences change even more than technologies and are the focus of innovation processes (Virta and Lowe 2016). The relationship between the media and the users is no longer unidirectional, but bidirectional and interactive. In addition, the user is a prosumer and participates in the creation of content. As users' needs and preferences evolve, media companies must adapt and develop new storytelling formats, distribution channels, business models and processes of users' interaction with content (Khajeheian and Friedrichsen 2017).

In this way, innovation in journalism should not be based exclusively on technology to produce the news, but also in the narratives and storytelling formats. As Miriam Hernanz, head of the *RTVE Lab*, argues, "it is not that journalism is innovative, but that narrative techniques are innovative. Journalism consists in telling stories with sources, facts and data. The way you must adapt your storytelling is what innovation means" (in García-Avilés et al. 2016: 187).

According to the *BBC The Future of News Report* (2015), audiovisual media face demanding challenges, such as the dissemination of content on multiple platforms, the commitment to interactive narratives and the need to manage Big Data. It is thus necessary to experiment with new languages, establish greater complicity between the media and the audience, incorporate journalists and presenters with whom young people feel identified and increase interaction with users (Kalogeropoulos et al. 2016). Some television producers are promoting initiatives to experiment with the news, but they are not able to change the newscasts, because these shows have become well established structures, where what always works is rewarded and anything new is rejected.

3 Research Objectives and Methodology

The purpose of our study is to establish how television news could be renovated, by collecting the opinions of the experts and their proposals, and by exploring audiovisual formats that are innovating in producing news content for younger audiences.

We posed two research questions:

RQ1: How television newscast could be renewed in the different stages of production, edition, presentation and distribution?
RQ2: What are the key elements of news formats which are innovating in storytelling and connecting with younger audiences?

An online questionnaire with professionals and experts was conducted in order to explore the current state of television news and to find out their opinion about how to innovate in designing, producing and distributing audiovisual news content. The sample was selected using a snowball method, among a list of 30 potential subjects, who were invited to take part in the survey. We received sixteen answers (a 54% response rate). All respondents currently work or have worked in television news; six work in public channels; three in commercial channels; five are university professors and two are consultants. Seven are female and nine are male, ranging from 33 to 65 years old.

The questionnaire consisted of three open ended questions: (1) Do you think TV news is in crisis? Why? (2) How can the traditional news format be reinvented? (3) Do you have any specific proposals to innovate in producing, editing, presenting or distributing television news? The answers were coded and analyzed, selecting the most relevant parts of the responses and identifying key themes and proposals.

We also analyzed a sample of innovative audio-visual news formats, developed by public and private broadcasters as well as digital native media. During a period of six months, we conducted research on databases and trade publications, and we identified different news formats. We then viewed samples of all them and selected six cases which were most relevant and innovative for the aim of this investigation. Those cases were then analyzed in order to identify the main elements in the format, exploring both their production process and their impact on the audience.

4 Experts' Perceptions About the Future of Newscasts

4.1 The Crisis of Television News

There is no unanimity about whether newscasts are facing a crisis. Most respondents agree that the traditional model of a self-contained programme which provides the news of the day at a specific time has become obsolete. Others argue that it is not so much a crisis but a transformation. In the words of a reporter, "it's not a crisis of TV news, but of those professionals who produce TV news in the same way it has always been done".

Several experts agree that television news is facing multiple crises in audience, content, competition and in its public service function. Given the multiplication of news available in the market, the traditional viewing experience is disappearing, and there is a less proactive public that seeks the news through television. "Producers still do not understand that the model of one direction communication is dead; they have not changed their way of doing the news for decades", says one academic.

As one journalist puts it, "newscasts continue in the twentieth century while viewers are in the twenty first century", consuming the news through social media in various platforms. "News formats keep reproducing a formula that worked for many decades but now, at the confluence between television, Internet and social media, is substantially modified", argues one interviewee. "TV news should have more influence on 'the how' and 'the why' of current events, since it cannot compete with the immediacy of social media", says one anchor.

According to some respondents, newscasts have long prioritized the function of entertaining without contextualizing the important issues. "The newscast is produced within a perverse dynamic: the golden minute; scarce international information; stories with little context…", explains a former TV journalist.

The credibility crisis is related to the diminishing of trust on television as a news medium, by its lack of independence from the political agents. In addition, some respondents argue that both public and commercial channels have been making concessions to populism and sensationalism, which have undermined their prestige as sources of serious information. Therefore, journalists need to be rigorous, checking the news and avoiding sensationalist content.

There is a consensus that quality, fact-checking, verification and contextualization are some of the values that news programs must incorporate. The concept of hard news is fleeting, while opinion is mixed with news and entertainment. Social media have become a popular source of news, but they should be used with journalistic criteria.

Nevertheless, several respondents argue that newscasts are not in crisis because most citizens still demand reliable and quality information. What has changed is the conception of traditional television, as a result of connectivity and the fragmentation of consumption in multiple devices and platforms. "Television is still the medium through which more people are informed and, above all, those who do not read the newspaper, who are increasingly more", says one reporter.

Several experts argue that television news will survive because the media face a permanent transformation, mainly due to a slow adaptation to the new technological paradigm of the Internet. The difficulty lies in maintaining current production, which still works among the older audience, and simultaneously creating something new for digital interactive platforms. "The problem is that most channels are reluctant to innovate. Most of the newscasts around the world resemble each other as drops of water", says one academic.

4.2   How to Reinvent the Traditional Newscast

Many respondents agree that producers should strengthen the basic journalistic values: news investigation and a vocation to tell the world through the eyes of its reporters. Newscasts should increase interpretation and analysis, offering viewers the key elements to understand what happens, allowing them to have their own opinion. Producers should also implement more innovation in storytelling and news supported by graphics, interactive tools and quality pictures.

From the technical perspective, newscasts are incorporating elements of Artificial Intelligence (AI), Augmented Reality (AR), Virtual Reality (VR), Immersive Journalism and 360° video, in which the viewer is directly connected to the news in real time. One interviewee says that in focus groups about using VR, people have been very reticent about VR in certain type of news: they do not want to 'get inside' of events such as wars, natural disasters or terrorist attacks. She recommends adopting AR formats, especially in sports news and weather forecasts. However, some professionals emphasize that technological advances are not the only way to renew the newscast.

Another common plea is more interaction with the public. "We must open the newscast to audience participation; it is the only way for many younger viewers to connect with a product which seems too traditional for them", states on expert. In this sense, the idea of participation must permeate the whole productive process: television journalists cannot remain alien to the conversation. It is useless to invite the public to take part if their contribution is not visible in any way. It is necessary to increase this interaction. For example, some channels are testing how journalists and viewers collaborate through news tips, access to sources, research support, etc.

There are also recurrent complaints that journalists and news editors do not listen to the audience. Instead, newscasters tend to focus on political conflicts, crime and bad news, to the detriment of those things what people care about. Some respondents argue that newscasts must be more open, giving 'the control of television' to the viewers, who could then choose the subjects that interest them. "It will be the viewers who make their own run, not us who impose the script of what they should watch", says one journalist.

Other experts suggest designing newscasts as a product in the value chain of a 'news factory', including news alerts, breaking news video, live narration in social media, as well as analytical reports and in-depth investigations. Each story should find the most appropriate format and platform to reach the public in an effective way. One academic also recommends creating listening mechanisms (surveys, scrutiny of the conversation in social networks, interaction).

4.3 Proposals for Innovating in Television News

The experts made many proposals, integrating elements of television news production, editing, presentation and distribution. We select the most relevant ones.

News production. Gatekeeping could be enhanced, giving coverage to the news that interests the audience and opening it to topics that have been covered from alternative sources or social media. Second screen tools could allow expanding the news on air and generating participatory dynamics which could be turned into content streams that can be easily visualized, such as surveys, viewers' comments, chats, etc.

Some experts recommend trying longer formats, more contextualized and in-depth that allow the viewer to choose how to move through the story. The current formats are too homogeneous and sometimes relegate important news to a mere voiceover narration. It is essential to manage the time of each news piece: not all should last the same. "We must edit and make news stories as autonomous pieces so that any image will have a text on it for the viewer to understand the story", argues one respondent.

Other proposals regarding the production of newscasts include:

- Access to automatic transcripts of scripts and videos;
- Producing content that could be consumed without presenters, reporting in a more proactive way and eliminating barriers with viewers;
- Checking sources and data automatically in real time, for example, during an interview. Also answering questions from the audience in real time.

News editing. Several respondents emphasize that the most urgent need is to recover credibility, offering relevant news that will once again fulfill the function for which the newscasts were created: helping the public to understand the world. "You must also apply criteria of rigor and quality in international or cultural news, which have been displaced by infotainment or impact stories", a TV reporter says.

Newscasts must diagnose social change, including topics of interest and reshaping the agenda, giving more weight to the human factor, making more visible those people who have something to say instead of those who are in 'the official agendas'. Producers need to understand that the world has changed, and the relevant issues are also different. One academic anticipated "shorter newscasts, combined with large programs where information, opinion and news are mixed".

Television should give voice to the whole of society: to politicians, only when their message is relevant and based on their representativeness; to the experts, indicating ascriptions and possible conflicts of interest; to unions, employers and social organizations that represent a relevant position; and to the concerns, demands, joys, and pains of the citizens. "This cannot be done by any algorithm, only with professional criteria, always fallible, but capable of listening, rectifying and giving voice to those who do not have it", adds one academic.

The journalist's job is to make the important things attractive. And the best way to do so is to promote audiovisual narration with sound storytelling that avoids turning the newscast into a succession of talking heads and unconnected videos. The content could be personalised in other platforms, but the newscasts should highlight the essential issues of public interest, with professional criteria. News and opinion converted into a show may be legitimate in other programs, but not in a newscast. But this does not mean that the news should be boring.

News presentation. Presenters should focus more on checking the news and providing context, explaining why it is important what they tell the audience. In order to innovate, the editorial team must strengthen its ability to present content of interest, focused on key issues and differentiated from other channels. They must offer something different to the digital platforms, exploiting the quality and potential of the image.

It is essential that the news presenter establishes a bond of trust with the viewers, telling stories with rigor and professionalism. One expert recommends improving the communication with viewers through a presenter in which to trust, with an attractive staging that "does not destroy the news story", giving priority to striking designs but that help to understand well what it is communicated. The presenter is described as "a prescriber in whose criteria the viewers trust". In a context of maximum speed and misinformation, this 'agreement of trust' with the journalist who presents the newscast is still valid.

The role of the analyst who can contextualize the most relevant news of the day should be implemented, at least in the nightly news. It often takes time to digest all that information that is aired, so it seems convenient to include more analysis and explanation.

News distribution. The 40- or 50-min long newscast makes sense for viewers who want a clear and precise summary of what has happened up to that moment. However, newscasts must be more interactive, offering a menu that will allow viewers to interact, vote, click on the graphic information, chronicles, etc. to provide with expanded information when the viewer demands it. This requires a broad team that works in a coordinated way with the rest of the news brand's windows (web, social networks, etc.) and develops transmedia contents.

Instagram, Twitter or Facebook have become fundamental tools in order to spread the content of newscasts. Newscasts should have open channels in the website, YouTube and social media, where reporters can explain the news provided in the broadcast. One model could be *BBC*'s *Open Source*, where viewers interact with the news and generate community. As one journalist puts it, "sometimes we make a live show for television and minutes later we make a Facebook Live. You can create labels for each piece specific to the channel and the news to share data, opinions, pictures and videos".

Producers should explore the possibilities of virtual reality and augmented reality for analytical and didactic narratives but avoiding pure spectacle. "I find innovation not so much in the spectacularization of the augmented reality or the graphics that overflow the screens, but in the clarity of the message, in its narrative values and in the confidence that arouses in the viewer", argues one producer.

5 International Cases of Innovative News Formats

In recent years, several audiovisual formats have emerged that seek to innovate in the approach of making and distributing news content. We have selected six international media formats which are innovating in different ways and are reaching some success in terms of audience, brand prestige or other aspects.

5.1 *Stay Tuned (NBC News, United States)*

It was launched on Snapchat by the commercial network *NBC News* in July 2017, as a format for mobile consumption. Subsequently, it is also broadcast on Instagram and YouTube. *Stay Tuned* uses pieces composed of close shots and few camera movements and titles, so that they can be understood without listening to the audio. The newscast lasts two minutes and is updated twice a day on weekdays and once a day on weekends. Presented by three journalists under 30 years old, the show combines what "the audience wants and needs to know", according to its executive producer (Digiday 2019). *Stay Tuned* offers breaking news and topics ranging from a youtuber's controversial statements to the humanitarian crises in Yemen, along with interviews with politicians, celebrities and activists.

The goal is to bring this new audience from Snapchat to its other platforms (Digiday 2019). *NBC News* has connected *Stay Tuned* to the rest of the network by having the hosts—Gadi Schwartz, Savannah Sellers and Lawrence Jackson—be involved in other programming, while also have *NBC News'* TV anchors participate in the Snapchat show. According to *NBC*, the newscast has an average of 35 million unique viewers per month and 75% are under 25 years old (Digiday 2019).

5.2 Hochkant (ARD, Germany)

The program, which began on October 1, 2016, is coordinated and financed by the Berlin regional broadcaster of public channel *ARD*. The production is run by a small editorial team in Berlin and three journalists abroad, who work together to develop pieces and topics. *Hochkant* uses a WhatsApp group to coordinate plans and discuss their most relevant issues and their coverage. Journalists do not have a fixed schedule, since they all cover different topics such as politics, entertainment and lifestyle. Between 6 and 9 in the morning they prepare a summary of the five most important news stories of the last 24 h; the rest of the day they cover breaking news.

The format is part of *Funk*, the social media network aimed at 14–29 year old launched by German public service broadcasters *ARD* and *ZDF* in 2016. *Hochkant* is anchored by three young presenters who call themselves 'snappers' and they generally switch between informational stills and short explanatory or commenting clips, depending on the urgency of the topic. The format is fit to provide a bit of context to people with short attention spans (Brachmann 2017). The presenters encourage their followers to interact, especially when they report on complicated or emotionally charged topics. They usually include pieces of User Generated Content (videos, pictures, comments…) and short quizzes in the shows.

5.3 ZIB100 (ORF, Austria)

Austrian public broadcaster *ORF* launched a video news summary via WhatsApp and Facebook to subscribed users. *ZIB100* offers two minutes of news pills at 17.25, when most citizens leave work. The programme, presented by the same anchors of the regular evening news, is produced in vertical video format and is subtitled. *ZIB100* can thus be watched on TV using a split screen and on a smartphone at ORF.at, WhatsApp, Facebook and Twitter. The format summarizes the main news about politics, economy, society and international affairs, in a professional and conventional style.

The news stories are based on short clips, which convey a message in a concise form through pictures, sound and subtitling. A video, for example, ideally is posted along with a textual description, enabling users to grasp the message without watching the video. Since its launch in April 2016, *ZIB100* has reached 110,000 subscribers and its WhatsApp distribution list has an opening rate between 60–70% (Reiter et al. 2017).

5.4 Outside Source (BBC, United Kingdom)

The format began as a radio show on *BBC World Service* in October 2013, created by journalist Ros Atkins, who since July 2017 presents the television edition. This format is broadcast simultaneously on *BBC World News* and *BBC News*. The idea was to bring the immediacy of the *BBC* newsroom, offering breaking news as they unfold, with the experience of the *BBC*'s global network of journalists. The presenter uses a touch screen to display graphics, images and content from social media, which help explain the context and background of the news in a visually appealing way.

According to Atkins "the advantage of the *Outside Source* screen is you can show developments better, so we can show things that come in text form in a more visual manner for the viewer, as opposed to traditional formats where you read the copy out, but you can't show it" (quoted in Reid 2015). They try to bring across all the very best bits of original journalism of the *BBC* newsgathering and they also seek out videos, pictures and social media comments from the audience. The programme has a team of journalists who verify and select user generated content.

5.5 NowThisNews (United States)

Founded both by former *Huffington Post* president Kenneth Lerer and former *Huffington Post* CEO Eric Hippeau, *NowThisNews* produces video content for social media since September 2012, with an editorial staff of 30 journalists that generate about 40 daily pieces, from breaking news to more elaborate analysis. *NowThisNews* delivers bite-sized videos, often aggregated from other sources, to a mostly millennial audience across various social platforms like Facebook, Twitter and Snapchat. Some are compilations of clips, while others are narrated by young presenters in a colloquial tone.

The pieces last between 15 and 120 s. While admitting that 15 s allowed a limited news release, managers expected viewers to have a knowledge base, so that their videos could add layers of meaning and context to a story and provide a fresh narrative style, departing from traditional television news (O'Donovan 2013). Viewers can consume a whole piece of content without ever having to go full screen or hear what is being said.

5.6   VICE News Tonight (VICE, United States)

Since October 2016, the show is broadcast Tuesday to Friday evenings on HBO in the United States and on YouTube, and it is produced by digital-only medium *VICE*, with a style very different from a standard newscast because it has no presenter in a studio. Instead, the pieces are introduced by a voiceover narration. Each programme begins with a quick-hit rundown of global events and a then a handful of feature snapshots of the world, strung together with slick graphics and music. A typical show uses a mix of voiceovers, graphics and video packages to dive into national and global news, technology, the environment, economics and pop culture. Its purpose is placing stories in context and understanding them.

Although intended for a US audience, *VICE News Tonight* includes many international news stories. The producers of the show claim that it reaches a younger audience than any cable news program in the country. According to his executive producer, his video inspirations are *Sesame Street* and *Saturday Night Live*, two multiformat playgrounds that get across their message in different ways, from monologues and sketches to song parodies and video clips (Zoglin 2017).

6   Conclusion: The Need to Experiment and Innovate

Most experts consulted in our study agree that television newscasts tend to the spectacularization of news content, with the use of high impact images and of extradiegetic music. This type of audiovisual language seeks that viewers empathize with the characters and the feelings in a story, reinforced by a fast, dynamic video shooting and editing style.

As some experts argue, innovation in newscasts essentially lies in the audiovisual narrative, integrating image and sound, telling the story with fluency, and holding viewers' trust with rigorous journalism. With the aid of graphics, augmented reality and social media, newscast content can be expanded. The relationship established by the presenters both with the elements of news staging and in their face to face relationship with the viewers can also be innovative.

However, innovation in television news should not only be technology driven. As many experts emphasize, the most innovative aspects are related to how news content is improved, the clarity of the message, its narrative values, its storytelling techniques, its power to engage viewers and the confidence that the journalists arouse in the public.

Our research shows that several selected formats are valuable examples of innovation in audiovisual storytelling: *Stay Tuned* (*NBC News*), *Hochkant* (*ARD*), *ZIB100* (*ORF*), *Outside Source* (*BBC News*), *NowThisNews* and *VICE News Tonight*. These formats do not provide all the solutions, but they offer opportunities for producers to risk, experiment and escape from their comfort zone, especially trying to reach viewers under 30 years old. However, if a format is made exclusively with the young audience in mind, there is a chance that the over-50 s, its current target audience, will abandon it. In the horizontal communication ecosystem of the Network Society, based on personalized content and interactive distribution, it seems impossible for a single format to reach a mass audience.

Even the most successful format will not be able to replicate the successes of the newscasts in the past. Given the variety of platforms available to obtain information, it is inevitable that many users feel that their needs are better served elsewhere. Perhaps, ultimately, the act of sitting down to watch television news becomes increasingly rare. Our findings suggest that young audiences must be taken seriously by news broadcasters for their long-term future. Young people demand content that is meaningful to them. Also, there is a need to reduce political and commercial influences to ensure independence in news coverage.

It seems naive to make predictions about the evolution of newscasts, given the amount of unexpected changes that lie ahead. Although it is likely that traditional newscasts will remain a relevant and valuable format for several years, there is much room for a contextualized, entertaining and visual daily product that helps make sense of an overwhelming volume of data and information. Television and the Internet, together with social media, seem optimal platforms for this format. The challenge is how to make it relevant, useful and satisfying the needs of users, without compromising the values of trust, transparency and credibility in a world of increasingly personalized and polarized media.

References

Barlovento Comunicación (2019) Análisis televisivo 2018. Retrieved from https://www.barloventocomunicacion.es/audiencias-anuales/analisis-televisivo-2018/ Accessed 25 Feb 2019

Brachmann E (2017) Hochkant — news coverage in portrait mode. Retrieved from https://medium.com/inside-the-news-media/hochkant-news-coverage-in-portrait-mode-d01778c667f7 Accessed 12 March 2019

British Broadcasting Corporation (BBC) (2015) Future of news. Retrieved from: https://www.bbc.co.uk/news/resources/idt-bbb9e158-4a1b-43c7-8b3b-9651938d4d6a Accessed 15 Feb 2019

Canter L (2018) It's not all cat videos: moving beyond legacy media and tackling the challenges of mapping news values on digital native websites. Digit JIsm 6(8):1101–1112

Digiday (2019) NBC News Snapchat show 'Stay Tuned' averages 25–35 m viewers. Retrieved from https://digiday.com/media/nbc-news-comscore-snapchat/ Accessed 5 Feb 2019

Drok N, Hermans L, Kats K (2018) Decoding youth DNA: The relationship between social engagement and news interest, news media use and news preferences of Dutch millennials. Journalism 19(5):699–717

García-Avilés JA, Carvajal M, Comín M (2016) Cómo innovar en periodismo. Entrevistas a 27 profesionales. Diego Marín, Murcia

García-Avilés JA, Carvajal-Prieto M, Arias F, De Lara-González A (2018a) How journalists innovate in the newsroom. Proposing a model of the diffusion of innovations in media outlets. J Media Innov 5(1):1–20

García-Avilés JA, Carvajal-Prieto M, De Lara-González A, Arias-Robles F (2018b) Developing an index of media innovation in a national market: the case of Spain. JIsm Stud 19(1):25–42

Kalogeropoulos A, Cherubini F, Newman N (2016) The future of online news video. Reuters institute for the study of journalism. Retrieved from: https://reutersinstitute.politics.ox.ac.uk/sites/default/files/research/files/The%2520Future%2520of%2520Online%2520News%2520Video.pdf Accessed 15 Feb 2019

Kalogeropoulos A, Nielsen RK (2018) Investing in online video news: a cross-national analysis of news organizations' enterprising approach to digital media. JIsm Stud 19(15):2207–2224

Khajeheian D, Friedrichsen M (2017) Innovation inventory as a source of creativity for interactive television. In: Friedrichsen M, Kamalipour Y (eds) Digital transformation in journalism and news media. Springer, Cham, pp 341–349

Newman N, Fletcher R, Kalogeropoulos A, Levy DA, Nielsen RK (2018) Reuters institute digital news report 2018. Reuters institute for the study of journalism. Retrieved from: https://reutersinstitute.politics.ox.ac.uk/sites/default/files/digital-news-report-2018.pdf Accessed 15 Feb 2019

Nielsen (2018) Q1 total audience report. Retrieved from https://www.nielsen.com/us/en/insights/reports/2018/q1-2018-total-audience-report.html Accessed 25 Feb 2019

Noguera Vivo JM (2018) You get what you give: Sharing as a new radical challenge for journalism. Commun Soc 31(4):123–141

O'Donovan K (2013) Instead of shoehorning it in, NowThis News is building video content that fits in where the audience lives. NiemanLab. Retrieved from: https://www.niemanlab.org/2013/09/instead-of-shoehorning-it-in-nowthis-news-is-building-video-content-that-fits-in-where-the-audience-lives/ Accessed 15 Feb 2019

Reid A (2015) How BBC outside source is forging a new, digital style of live video news. Journalism.co.uk. Retrieved from https://www.journalism.co.uk/news/how-bbc-outside-source-is-forging-a-new-digital-style-of-live-video-news/s2/a565342/ Accessed 18 Mar 2019

Reiter G, Gonser N, Grammel M, Gründl J (2017) Young audiences and their valuation of public service media. In: Lowe GF, Van den Bulck H, Donders K (eds) Public service media in the networked society. Nordicom, Gotemburg, pp 211–226

Terán B (2017) El fin del Telediario: así será la revolución de los (caducos) informativos de las cadenas. Lainformacion.com. Retrieved from https://www.lainformacion.com/opinion/borja-teran/el-fin-del-telediario-asi-sera-la-revolucion-de-los-caducos-informativos-de-las-cadenas/6338302/ Accessed 5 Jan 2019

Virta S, Lowe GF (2016) Crossing boundaries for innovation: content development for PSM at YLE. In: Lowe GF, Yamamoto N (eds) RIPE@ 2015. Crossing borders and boundaries in public service media. Nordicom, Gotemburg, pp 229–246

Zoglin R (2017) Inside vice's effort to reinvent the evening news, the bridge. Retrieved from https://thebridgebk.com/vice-news-tonight-reinvents-evening-news/ Accessed 15 Mar 2019

Jose A. García-Avilés Full Professor of Journalism at Miguel Hernández University (Spain), where he lectures at the Master Programme in Journalism Innovation. He is Bachelor of Arts (National University of Ireland) and Ph.D. in Communication (University of Navarra). He was visiting scholar at Columbia University and has carried out comparative research on journalism innovation and media transformation. He is director of the Communication Research Group GICOV and founder of InnovaMedia.Net, a network of researchers on journalism innovation.

Mediamorphosis of Participation on Television: The News Programmes

Ana González-Neira and Natalia Quintas-Froufe

Abstract Convergence processes have imposed transformation dynamics to the so-called traditional media. Through the digitalization of television, the ruling media, there has been a huge evolution that has altered its usual forms of production and consumption. In this paper, we reflect upon the new forms of participation of the audience in television fostered by these processes of convergence. Subsequently, the existing windows of participation are analysed in scarcely studied format, such as news programmes. The results of this research will allow to confirm whether these informational television spaces fit in these new forms of television.

Keywords Television · Audience · Information · Participation

1 Contextualization

The transformation of the ruling media, the television, in the past five years has staggered some of the principles of its operation that were still active since its origins. The ATAWAD key principle (anytime, anywhere, any device) breaks with the existing concept of television since the 40s. As a result of these changes the liquid television emerges (Quintas-Froufe and González-Neira 2016) based on Bauman's ideas (2006). In this new reality it is very difficult to organise and delimit the flows between actors present in the communicational process of television. All these changes have promoted, at the same time, an empowerment of audiences, a re-assessment and re-definition of its role in communication.

The scientific literature about transformations of television is quite profuse (Strangelove 2015; de Valck and Teurlings 2013; Turner and Tay 2009; Lotz 2007; Spigel and Olsson 2004). In this chapter we aim to analyse these evolutions from the perspective of the changes in the possibilities of participation of the audience, in

A. González-Neira (✉) · N. Quintas-Froufe
Department of Sociology and Communication Sciences, Universidade Da Coruña, A Coruña, Spain
e-mail: ana.gneira@udc.es

N. Quintas-Froufe
e-mail: n.quintas.froufe@udc.es

© Springer Nature Switzerland AG 2020
J. Vázquez-Herrero et al. (eds.), *Journalistic Metamorphosis*,
Studies in Big Data 70, https://doi.org/10.1007/978-3-030-36315-4_12

short, to know to what extent the public has been granted a greater power. After an explanation about this evolution, there will be an analysis of the specific case of the possibilities of interaction offered by news programmes broadcasted by the Spanish general-interest networks. As Livingstone (2019) states, we believe it is necessary to contextualise the audience phenomena in the society they take place instead of only offering unconnected data. Therefore, there is offered a research based on the triangulation of three key concepts: television, participation and news programmes. After a brief description about the technical evolution of television and participation, the study delves into the possibilities present in the informational format of news programmes.

Like radio, when the television was born it tried to copy the genres that were successfully established in the radiophonic medium. At first, there was more preference for entertainment (contests, series) since the technological limitations hindered the fast broadcast of images that illustrated the news. Thus, it was demonstrated during the Pearl Harbor bombing with the nine-hour coverage of the experimental broadcasting station of CBS: WCBW. However, by the late 40s, the informative appointments with the television became fixed through two examples: the first *Journal Televisee* of the French public channel in 1949 and *Newsreel with John Cameron Swayze* from *NBC*.

These two examples mark the development of the two informational models, the European, which is more institutional and the North American, whereas the host has a greater leading role. From its beginnings, this rather rigid format has undergone some modifications (sets, duration, etc.) but often mainly driven by technological reasons (García de Castro 2014).

2 Transformation of the Medium

2.1 Metamorphosis of the Television

The digitalization processes that have impacted the different traditional media in the past couple of decades have transformed their nature. The media ecosystem has mutated and there is no longer any species resembling those that existed twenty years ago. The digitalization processes and media convergence have altered the nature of the press, radio and television, at the same time they have promoted the emergence of other new species like social networks or native digital media. In this paper there will be a progress in the case of television, reviewing this transformation process experienced in these past years and assessing how it has influenced in its relationships with the audience.

The technological changes have delimited the different stages of the history of television by allowing to acquire other functions and new experiences of use for the spectator. Said changes have been accompanied by innovative entrepreneurial and sociological processes. Let's think about the progressive incorporation of colour, the presence of mobile devices and satellites up until reaching the boom of cable TV back in the 80s. But undoubtedly, the digitalization of television services and the incorporation of the Internet represented a radical change in the concept of television and its connection with the spectator. In order to facilitate the explanatory coherence, we follow the stages set by Johnson (2019) who indicates that television has shifted across the Broadcast era (1930s–70s), Cable/Satellite era (1970s–90s), Digital era (1990s–2000s) and the Internet era (2010s+). In this last stage there would be the emergence and huge penetration of mobile devices, increasingly potent, as well as the expansion of social networks that interact with media. All this process would coincide in a subsequent stage to the neo-television coined by Eco and that Scolari (2008) denominates hyper-television due to its hypertextual nature.

Considering the object of study in this paper, we must also be interested in the concept of interactive television and its evolution, in such a way that the possibilities offered to the audience in the current television can be studied. Except for some occasional experiences, the development of interactive television started by the end of the seventies, along with the cable television. Qube, Viewtron or Full Service Network (FSN) were some of the initiatives born in the last decades of the 20th century century. For Marinelli (2015), the failure of all these experiences was due to the precarious interfaces used and the lack of preparation from the public, still hardly acquainted to the interactive dialogue of the PC or the web browser. After the digitalization and the improvement in the connection infrastructures, the interactive television would intensify. PVR Systems (Personal Video Recorder) in the set top box would facilitate some of these experiences of interactivity in a growing contest for Smart TVs.

In this area, the new broadcasting agents such as Netflix, HBO or Amazon Prime have undoubtedly entailed a burst of interactive television since its on-demand operation involves interactivity. To select, record, resume the viewing in the same spot it was left days earlier, to rewind, etc. are some of the possibilities used today by any spectator while viewing contents coming from these new actors.

2.2 The New Screens

After a first moment of settlement, the convergence of television on mobile devices has multiplied the fruition possibilities of the public. This interruption has undoubtedly encouraged the breach of the viewing-indoor axis. In the same way the transistor made the radio to go outdoors in the 50s, smartphones and tablets allow the consumption of television in motion. The possibilities of consumption multiply and at the same time there is a transformation in the type of audience. Like Athique points out about with the new technological incorporations "our conceptualization of the media audience becomes at one radically individuated and densely interconnected" (Athique 2016: 61).

On the other hand, these mobile devices have caused an increasingly individualised consumption (with the subsequent hyper-fragmentation of audiences) before the many windows for spreading television content. The viewing through mobile devices produces a breach in space and also in the time of viewing. Now, the audience has the capacity to become its own programmer. The relationships of power between the broadcaster and the audience have balanced and it is the spectator who decides his or her own television diet, when, where and what is consumed. "Mobile viewing on handheld devices allows for 'place-shifting' as a complement to time shifting. All these, promise to make the viewer into the programmer" (Newman and Levine 2012: 131).

2.3   Changes in the User Experience

We cannot forget the improvements of the television signal in the distribution systems. The quality of the image has improved noticeably. The screens have become progressively flatter (HD, 4K, etc.) and larger, following the trail of cinema. In addition, the arrival of the Internet introduced Smart TVs and turned television devices into giant computer screens (Johnson 2019: 14). The implantation of the new television devices was quite fast in countries like USA. In 2009, a third of homes in the United States owned an HDTV set. Slowly, there was heading towards a sort of view that imitated the experience in the movie theatre with surround sound system while the aspect ratio of television sets shifted from 4:3 to 16:9.

However, we are positioned in a moment where there is a maturity of the public regarding the use of interfaces and the management of the Internet that improve the user experience. As Marinelli indicates, "It is the same television medium that, passing from broadcastig to flow generated user (Uricchio 2010) aims almost naturally to incorporate interactivity as the driving principle of the viewing" (2015: 280). It is the user who decides what, when and how a television content is consumed and interacted with. All this process is reinforced by the multi-screening practices favoured by the second screen devices and the social media. Therefore, creating a more complex and complete viewing experience.

2.4   Changes in the Sociological Conception of Television

We cannot limit to these technological transformations that have motivated the onset of changes regarding television of the past years. As indicated by Amanda Lotz (2016) to talk about the medium, we do not only need to consider the technological contributions but also its textual features, industrial practices, audience behaviours and cultural meanings. It is important to analyse the changes in the meaning of current television. In these last years, there has produced a cultural transformation of television legitimation. It has revalued from the cultural parameters and it turned from being called the 'idiot box' by many to the platform of cultural products comparable to masterpieces of literature or cinema (Newman and Levin 2012).

Likewise, we have witnessed a remarkable improvement in the usability and quality of the image since 2014. At homes, television occupies a prominent place in living rooms once more, considering the arrival of Smart TV. It even appears as a decorative element, hanging from the walls as if it were another picture or a photograph. From the second half of the first decade of the 21st century, television devices get a room dominating size. Like Newman and Levine say, "Along with other digital devices used to select content to view on them, flat panel TVs have been essential in creating a sense of television's renewal and improvements as a technology and as medium" (2012: 102).

3 Metamorphosis of Participation

Participation has been one of the leading role concepts in communication studies in the past years. Literature about its definition (Bergillos 2015) and its evolution (McElroy 2019) is quite abundant and has been approached from different perspectives. In this paper, we pay attention to the dependency there is between participation on media and technological possibilities. We consider participation as a type of interactivity, with the characteristics that García Avilés points out, "as the feedback which the broadcasters provide through a combination of traditional systems and new technologies" (2012: 430). Namely, the study of the possibilities that the technology has offered for television networks to provide interaction with the audience and the harnessing of these possibilities.

However, it is necessary to allude another concept that is usually connected and that has been the focus of the work of many researchers: interactivity. McElroy (2019), Jenkins (2008), León and García Avilés (2008) and Rost (2004, 2006) have focused their researches on interactivity. The latter made an interesting difference between selective and communicative interactivity. While the former refers to the capacity of the user to choose the contents to be consumed, the communicative interactivity entails further level of intensity since there is granting of a greater protagonism when offering the possibility to produce, comment or modify contents.

It is deemed convenient to carry out a small review about the possibilities that television has offered throughout its history. Despite the fact they are not identical realities, we must focus on the experience of interactive television. Since the origins of television, these realities have been very limited and with scarce success until the arrival of the new century. We started with the cartoons *Winky Dink and you* broadcasted in the USA during the 50s, where children participated in the narrative development of the story by sticking some slides to the screen through static electricity. In the 70s, with the onset of the cable/satellite era indicated by Johnson (2019) there were developed in USA and Japan some more complete experiments but without full success (León 2012) except for the remote control. It is precisely this device, together with videocassette recorders (VCR) the ones that granted greater power to audiences, since there was the possibility to decide what is consumed (in a more comfortable way using the remote control) and when (breaking the line of simultaneity between broadcasted and consumed content). We must not forget that it is in the 70s when the television turns into the interface for not purely television-related activities, as per the concept known until then. There were incorporated add-on devices that improved the user experience:

> set-top boxes aggregated programmes in new ways; VHS players enabled television programmes to be recorded, sold and/or rent as products; games consoles turned the television set into a screen for playing as well as viewing; and remote controls enabled viewers to change channel from the comfort of their sofas (Johnson 2019: 7).

Considering the concept of participation, the possibilities prior to the arrival of the Internet were very limited and with scarce success. In the beginning, the only forms of participation for the audience were face-to-face, namely, being part of the public during recordings of TV programmes in theatres or sports pavilions. The audience had very limited resources available for interacting with the broadcaster. The relationships of power between the sender and the sender were still quite uneven, therefore participation was almost testimonial. The letters to the director, sending postcards/coupons, phone calls and later SMS texting would be the only mediated windows of participation in television programmes, generally of entertainment.

With the digitalization the number of interactive experiences in television intensified, at the same time the nature of these windows of participation changed. The novelty that the Internet provided is that while previous possibilities of participation such as letters/postcards or phone calls were always held between receiver-sender; today the communication can expand towards a greater collectivity instead, towards a community, fostering the concept of social television. Said concept refers to the type of interactive television that results from the technological convergence where spectators participate (by writing comments, reading, etc.), in the contents through social networks or other channels and thus, use second screen devices (González-Neira and Quintas-Froufe 2015). There is shifting from the bidirectional dialogue to a conversation where multiple anonymous actors participate.

The landing of the Internet represented a booster in the progresses of interactive television. It needed to adapt to the arrival of this new medium and even mimic some of its defining features, in a way. The interactivity inherent to the Internet, is one of them. No interactivity, no Internet. In a convergence process, television enriches when it is spread of interactivity and participation strategies coming from the Internet; it appropriates and incorporates them to its operation. In such a way, the ruling medium acquired greater possibilities of communication and contact with the audience. Furthermore, from the narrative perspective, the Internet also favoured the expansion of the media text and the spreading of transmedia extensions.

In this change in the possibilities of participation, the irruption of mobile devices mentioned earlier has also influenced notably. We coincide with Newman and Levine when pointing out that through the study of usability and design of interfaces "The television viewer's rebirth as the television user is one outcome of the convergence of

TV with computers" (2012: 150). Again, the technological changes like the development of Wi-Fi, 4G, cloud computing, fibre optic cables, and also the Internet Protocol Television (IPTV) and over the top (OTT) are the baseline of this increase of participation possibilities. Likewise, cultural and social changes have also facilitated the installation of these participation dynamics. The current spectator is immersed in the participatory culture (Jenkins and Ito 2015). Currently, the user has assumed interactivity as something intrinsic to his or her digital life, therefore participation appears as an essential element of the new mediatic reality. The user demands that aforementioned interactivity, with increasing levels of specialization, is present in the media he or she consumes. The spectator of the current television has a profile that is active, multimedia, multiplatform, in continuous contact with the Internet and with information technologies and used to online viewing (García García et al 2012). Therefore, considering these consumption practices, they must satisfy an increasingly demanding user experience.

The synergies provided by websites and social media facilitated the introduction of more complete transmedia strategies that leveraged the possibilities of participation from the audience. Perhaps one of the most recent examples of this sort of experiences of participations was the web series *Si fueras tú* by *TVE* where through voting on Twitter and Facebook, spectators decided how every chapter would end, thus directly participating in the narrative development of the programme.

As Andò (2018) and Napoli (2011) indicate, the different participation practices, like writing comments about a text or to share it, produce an increase of audiences since there is an increase of the possibilities of access to that comment through different platforms (social networks, apps, websites, etc.). In addition, like McElroy explains, "interactivity causes higher levels of audience engagement, it is a tool that media producers can use to tout and sell their audience to advertisers" (2019: 452). In a business model based on advertising, like that of general-interest networks, this life expectancy of a television space beyond its original broadcasting schedule and the increase of customer loyalty are crucial to reach good ratings. However, regardless of the implantation of these new forms of participation, the harness by the audience is still very low. Like Bergillos points out, "it is an industry with vertical structures in which participation in the medium is practically non-existent" (2017: 89).

These technological innovations have reactivated the theory of active audiences (Livingstone 2004) opening new and interesting fields of study. The possibilities that these transformations offer to audience research are an interesting opportunity "to contribute to theoretical and methodological development across the wider field of media and communication research" (Jensen 2019: 151).

4 Participation Windows in News Programmes

Considering the mediatic context explained and the evolution of the television mentioned, we deem necessary to highlight participatory elements coming from the broadcasting and digital platforms. As observed along this chapter, there are several studies that have focused on the analysis of participation and television. Besides the already mentioned, we cannot forget Murschetz and Schlütz (2018) or McElroy (2019). However, there hardly is scientific literature that jointly comprises the three pillars on which this research is based: participation, television, news programmes (Cazajeira 2015; Sa 2015; Lago and Campos 2016; Herrero, 2016). The studies of Steensen (2014) and Bird (2011) analyse the participation in the field of journalism even though they do not circumscribe to television. In the Spanish case, there outstand the studies of García Avilés et al. (2019), Bergillos (2017) and García Avilés (2011, 2012). We agree with Bird (2011) about the idea that researches regarding journalism on television tend to be analysed from the perspective of the producer rather than the audience's. On the other hand, it is worth reminding that studies about the possibilities offered to the audience on television have mainly focused on entertaining formats (contests and fiction) notably influenced by the trend of cultural studies.

The object of this research is the study of participation possibilities offered by news programmes in general-interest networks in Spain. This decision is justified considering the quite scarce literature about news programmes, television and participation. The news programmes is one of the oldest, solid and most disseminated television formats in the traditional television networks. Since its birth by the late 40s, it has barely undergone variations through these decades. In Spain, it still appears as one of the most viewed spaces daily (Rubio et al. 2018).

Up until now, the creation of a fandom movement around this format broadcasted in Spain has not been confirmed. In all of them, there is option for journalists that do not personalise the news programmes, compared to what is done in other countries or in other informative programmes of different format like the USA. When lacking this personalization and opting for yielding the protagonism to information, customer loyalty becomes more difficult.

To conduct this research there was selected a sample composed of the news programmes broadcasted during the prime time by Spanish general-interest networks (*La 1, La 2, Antena 3, Telecinco, La Sexta*) as well as the only thematic channel about news programmes: *24horas* belonging to the *RTVE* group.

After a first stage of literature review, an analysis card was elaborated. In order to do so, the studies of Bergillos (2015) and García Avilés (2011) were considered and adapted to the format, object of analysis. The analysis table comprises the participation possibilities (García Avilés 2011) about providing feedback from the public (to react, comment, complete surveys), distribution of contents from users (to share) and production (sending pictures, information and suggestions). The participation windows offered by the different networks that compose the analysis card include the audience as recipient of a television message and not as protagonist of the information offered in news programmes, namely, there was chosen a type of mediated participation excluding the face-to-face modality. Likewise, there are considered the participation windows accessible to the population in its entirety, and not only to those individuals who live in the centres of production. Based on the mediated participation, in the analysis table elaborated, the digital presence of each one of these news programmes was considered (website, Twitter, Facebook, YouTube, Instagram and Spotify).[1]

The study of these possibilities of participation was done after viewing the different news programmes object of study. The invitations to aforementioned participation are limited mostly to be included in an overprinted manner at some point during its operation (generally in the end), the Twitter account (*Telediario de La 1, La 2 noticias, Antena 3, La Sexta Noticias*), the website (*Antena 3 Noticias, Informativos Telecinco, La Sexta Noticias*), on Facebook (*Antena 3 Noticias, La Sexta Noticias*), on Instagram (*Antena 3 Noticias*). Only in *La 2 Noticias* appears the hashtag of the day (for instance #La2N15a) in a corner of the screen to serve as backbone for conversations on Twitter. Never, any of the hosts addresses the public inviting them to comment, participate, send messages (unlike entertainment programmes). With these overprints, the networks aim that the spectator keeps following the information on their digital extensions rather than promoting interaction or participation. Namely, their purpose is extending the news programmes brand beyond television without paying much attention to interaction possibilities (Table 1).

After the analysis, it is confirmed that all of them have a specific website space available that is used as extension of what is broadcasted and an update of spread news. Regarding social media, it is perceived that *Antena 3* is the network with greatest presence (Twitter, YouTube Facebook, Instagram and Spotify), compared to the contempt shown by *La 1* which is only present on Twitter. This presence is relevant because a great part of the possibilities of participation precisely come from the digital derivations of news programmes and the instruments provided by social media. Namely, the reaction towards a content will be produced whereas that channel, the recipient of information coming from the public, is present. Hence, the possibilities of participation of the 'reaction' mode are minimum in the *Telediarios de La 1* (since it is only present on Twitter) compared to the possibilities of *Antena 3 Noticias*. On the other hand, this analysis concludes that none of the studied programmes elaborates surveys for spectators, despite social networking sites like Twitter, Facebook or Instagram allow to do so.

Table 1 Presence of news programmes on the different platforms. Own elaboration

Channels	Web	Twitter	YouTube	Facebook	Instagram	Spotify
Telediario TVE	✓	✓	✗	✗	✗	✗
La 2 noticias	✓	✓	✗	✓	✓	✗
24 horas	✓	✓	✗	✓	✓	✗
Antena 3 noticias	✓	✓	✓	✓	✓	✓
Informativos Telecinco	✓	✓	✗	✓	✓	✗
La sexta noticias	✓	✓	✓	✓	✗	✓

---

[1] The monitoring was conducted in March and April 2019.

It has been confirmed that the action of sharing, compared to the participation as form of distribution, is the most usual in all networks; therefore there is a clear option for the spreading of contents to third parties, the extension of information and the creation of communities around these spaces.

The research performed shows that the action of 'comment' is only possible through social networks. Namely, none of the websites of the news programmes allow the Internet user to express his or her opinions about a content (unlike the case of written online press). This possibility is quite related to 'sending images, information or suggestions', an option that only the three spaces of the *RTVE* group include. In other words, the private broadcasters do not have any space on their websites available to comment or send information. The possibilities of access of the citizen journalism are remarkably limited in this sense.

Twitter has turned into one of the most common channels of participation for the audience. For many years, the practice of issuing comments in a television programme while watching has been established as something usual among spectators who comment, send information and distribute related content to third parties. Like Bergillos points out, the social networks have turned into "essential places for managing social conversations linked to television scheduling" (2017: 100). Even though it has been confirmed that the entertainment programmes (talent, reality shows) are the ones reaching a greater index of social audience, the informative spaces slowly sneak into this sort of classifications. Despite their structure does not promote the fandom phenomenon that strongly influences the social audience, some of the news programmes analysed appear as the most tweeted programmes of the month. Since 2016, *Antena 3 Noticias* is the space of this format with greater success. That year became the fourth most tweeted programme (behind the different versions of *Gran Hermano* and *Eurovision*) while in 2018 it was positioned seventh behind *Gran Hermano*, *Operación Triunfo*, *Eurovisión*, *Supervivientes* and *Sálvame*. In the 11th position, there is *La Sexta Noticias* (Kantar Media 2019). Since then, it has maintained the leadership among the news programmes format in social audience, followed by far by *La Sexta Noticias*. It is undoubtful that in said success the number of followers of every account has an impact because there is a parallelism between the number of followers and the success in social audience. On April 30, 2019 the profile @antena3noticias had more than one million seven hundred thousand followers, @Lasextanoticias surpassed the one million one hundred thousand, while @infomativost5 barely exceeded six hundred thousand. The most noteworthy case was that of @telediario_tve that did not reach two hundred thousand, a very low number of this public channel.

5   Conclusions

The technological progresses have boosted the development of the participation windows of the audience in the current television. In a context of convergence, the television has appropriated elements coming from the Internet to increase the points of contact with the public. In the case of news programmes, these possibilities contributed by the digital context have introduced greater opportunities of interaction, practically non-existent until then in this format. In the model of Spanish news programmes, the mediated participation routes have been displaced to the digital field, unlike what happens in other contents.

The research performed demonstrates that, despite being a quite rigid format, news programmes do not make the most of all the existing possibilities to offer their audience bidirectional communication channels. From television sets, hosts do not invite the audience to participate in the different windows available and limit to show the overprinted social network accounts or the programme's website, with a merely informational purpose, of extension of the content, instead of promoting the dialogue with the audience.

There outstands the presence of news programmes of the channel *Antena 3* since it is present in all platforms analysed, unlike the news programmes of *La 1*, that only have Twitter available as well as a website of its own.

The possibilities of participation, of achieving feedback from the public, are supported on structures external to television channels like social networks (Twitter, Facebook, Instagram, etc.) and not on the websites of this kind of news programmes. None of the private broadcasters have any channels to comment, send images or information available on their websites.

These results conclude that, in the Spanish case, the harnessing of one of the forms of participation that allow to capture and foster customer loyalty of new spectators is being wasted, a form that also allows to create a community beyond the television screen and improve the user experience. However, in this paper the rigid format of the news programmes has been considered, which hinders the implantation of other possibilities of participation present in other formats such as fiction or contests.

It is worth mentioning that this paper is a first approach towards the possibilities of participation, without considering other studies with more complete taxonomies. Therefore, this research leaves other research lines opened, such as the analysis of the use of each one of these options in every space analysed; namely, to conduct a study from the audience perspective, about the actual use of these diverse possibilities and a comparison with the behaviour in other European countries. Likewise, there could be progress in the study of content of news programmes to confirm whether it suffers some sort of alteration to pull towards participation.

Acknowledgements We thank the student Laura Cotelo for her collaboration in this research.

References

Athique A (2016) Transnational audiencies. Media reception on a global scale. Polity Press, Cambridge

Andò R (2018) Esperienze televisive. Schermi, rituali e pratiche delle audience connesse. In: Andó R, Marinelli A (eds) Television(s). Come cambia l'esperienza televisiva tra tecnologie convergenti e pratiche social. Guerini, Milan, p 161–198

Bauman Z (2006) Modernidad líquida. Fondo de Cultura Económica, Buenos Aires

Bergillos I (2015) Participación de la audiencia y televisión en la era digital. Propuesta de análisis y evolución de las invitaciones a la participación en la tdt y en otras plataformas. PhD dissertation, Universidad Autónoma de Barcelona

Bergillos I (2017) Invitaciones a la participación de la audiencia a través de la televisión en España: análisis del prime time de los canales generalistas en 2010 y 2014. Quaderns del CAC 43:91–104

Bird SE (2011) Seeking the audience for news: Response, news talk, and everyday practices. In: Nightingale V (ed) The handbook of media audiences. Wiley, Chichester, pp 489–508

Cazajeira PE (2015) A Audiência ubíqua do telejornalismo nas redes sociais. Serra P, Sá S, Souza Filho W (orgs) A televisão ubíqua. Livros LabCom, Covilhã, pp 169–190

de Valck M, Teurlings J (2013) After the break: television theory today. Amsterdam University Press, Amsterdam

García Avilés JA (2011) Dimensiones y tipología de las actividades de participación de la audiencia en la televisión pública. Ámbitos. Revista Internacional de Comunicación 20:175–195

García Avilés JA (2012) Roles of audience participation in multiplatform television: From fans and consumers, to collaborators and activists. Participations. Journal of Audience and Reception Studies 9(2):429–447

García Avilés J, González Mesa I, García Ortega A et al (2019) La crisis del informativo televisivo. Cómo innovar en los formatos audiovisuales. Compobell, Murcia

García de Castro M (2014) Información audiovisual en el entorno digital. Tecnos, Madrid

García García AL et al (2012) Nuevas fórmulas de producción audiovisual atendiendo a criterios de interactividad. In: León B (ed) La televisión ante el desafío de internet. Comunicación Social Ediciones y Publicaciones, Salamanca, pp 122–129

González-Neira A, Quintas-Froufe N (2015) Revisión del concepto de televisión social y sus audiencias. In: Quintas-Froufe N, González-Neira A (eds) La participación de la audiencia en la televisión: de la audiencia activa a la social. Asociación para la Investigación de Medios de Comunicación, Madrid, pp 13–26

Herrero M (2016) Análisis de la audiencia social en la programación informativa de prime time. El caso de Atresmedia. In: Saavedra M, Rodríguez L (eds) Audiencia social. Estrategias de comunicación para medios y marcas. Síntesis, Madrid, pp 73–88

Jenkins H (2008) Convergence culture: la cultura de la convergencia de los medios de comunicación. Paidós, Barcelona

Jenkins H, Ito M (2015) Participatory culture in a networked era: A conversation on youth, learning, commerce, and politics. Polity Press, Cambridge

Jensen KB (2019) The Double Hermeneutics of Audience Research. Telev New Media 20(2):142–154

Johnson C (2019) Online TV. Routledge, New York

Kantar Media (2019) Anuario Social TV 2018. Retrieved from https://www.kantarmedia.com/es/blog-y-recursos/data-lab/anuario-social-tv-2018. Accessed 15 May 2019

Lago D, Campos F (2016) Las redes sociales como foco de interactividad del ciberperiodismo de pantalla: el uso de Twitter en los informativos de televisión. In: VII Congreso internacional de ciberperiodismo y web 2.0 Nuevos perfiles y audiencias para una democracia participative. Universidad del País Vasco, Bilbao, pp 222–239

León B (2012) La televisión frente a internet. Unas historia por escribir. In: Leon, B (coord) La television ante el desafío de internet. Comunicación Social, Salamanca, pp 19–29

León B, García Avilés JA (2008) La visión de los productores sobre la televisión interactiva: el final de la utopía. Comunicación y Sociedad 21:7–24

Livingstone S (2004) The challenge of changing audiences: Or, what is the audience researcher to do in the age of the Internet? Eur J Commun 19(1):75–86

Livingstone S (2019) Audiences in an age of datafication: critical questions for media research. Telev New Media 20(2):170–183

Lotz AD (2016). The paradigmatic evolution of US television and the emergence of internet-distributed television. Revista Icono14 14(2):122–142

Lotz A (2007) The television will be revolutionized. New York University Press, New York
Marinelli A (2015) L'interattività della televisione. In: Arcagni S (ed) Da innovazione mai real-izzata a pratica quotidiana nel networked media space. In I Media digitali e l'interazione uomo-macchina, Aracne, Roma, pp 275–304
McElroy BP (2019) Experimenting with interaction: TV news efforts to invite audiences into the broadcast and their effects on gatekeeping. Convergence 25(3):449–465
Murschetz PC, Schlütz D (2018) Big data and television broadcasting. A critical reflection on big data's surge to become a new techno-economic paradigm and its impacts on the concept of the "addressable audience". Fonseca, Journal of Communication 17(2):23–38
Napoli PM (2011) Audience evolution: new technologies and the transformation of media audiences. Columbia University Press, New York
Newman MZ, Levine E (2012) Legitimating television: Media convergence and cultural status. Routledge, New York
Quintas-Froufe N, González-Neira A (2016) Consumo televisivo y su medición en España: camino hacia las audiencias híbridas. El profesional de la información 25(3):376–383
Rost A (2004) Pero, ¿de qué hablamos cuando hablamos de interactividad? Center for Civic Journalism 2:1–16
Rost A (2006) La interactividad en el periódico digital. PhD dissertation, Universidad Autónoma de Barcelona
Rubio LG et al (2018) La televisión informativa en el entorno digital: análisis de las ediciones diarias de las cadenas generalistas líderes de audiencia en España. Estudios sobre el mensaje periodístico 24(1):193–212
Sa S (2015) O espectador em alta definiçao. In: Serra P et al (eds) A televisão ubíqua. Livros LabCom, Covilhã, pp 145–168
Scolari CA (2008) Hacia la hipertelevisión: los primeros síntomas de una nueva configuración del dispositivo televisivo. Diálogos de la comunicación 77
Spigel L, Olsson J (2004) Television after TV: essays on a medium in transition. Duke University Press, London
Steensen S (2014) Conversing the audience: a methodological exploration of how conversation analysis can contribute to the analysis of interactive journalism. New Media Soc 16(8):1197–1213
Strangelove M (2015) Post-TV: Piracy, cord-cutting, and the future of television. University of Toronto Press, Toronto
Turner G, Tay J (2009) Television studies after TV: understanding television in the post-broadcast era. Routledge, London
Uricchio W (2010) TV as time machine: television's changing heterochronic regimes and the production of history. Relocating television. Routledge, London, pp 49–62

Ana González-Neira Associate Professor in Universidade da Coruña. She is Ph.D. and graduate in Journalism and Political Sciences. Her research lines focus on the history of journalism, the study of the new audiences and mediatic consumptions. As a result of this study, she has published several research papers and two books. She has been visiting professor in Universidad Autonóma de México, Università della Sapienza of Rome and Universitá Cattolica of Milan.

Natalia Quintas-Froufe Full Professor in Universidade da Coruña. Graduate in Advertising and Public Relations and Ph.D. by Universidad de Vigo. Currently, her main research lines focus on the new advertising formats and the study of new television audiences. She has published several papers on national and international journals about these subjects. She has performed research stays in Università Cattolica del Sacro Cuore, Universidade Católica Portuguesa and Católica Pontificia Universidad de Chile.

Value and Intelligence of Business Models in Journalism

Manuel Goyanes, Marta Rodríguez-Castro and Francisco Campos-Freire

Abstract Added value and intelligence, both in strategies as well as in operating and technical applications, are essential to reinforce trust and sustainability of business models of old and new media. This chapter addresses these two areas in the main conceptual and structural elements of business models in press, audiovisual and digital platforms, taking into account the diversity of expressions: payment systems of physical, digital and mobile contents, premium, open access, bundle, native and programmatic advertising, sponsorship, bartering, membership, crowdfunding, foundations, events, YouTubers, influencers, data, augmented reality and artificial intelligence, gamification, Internet of things and blockchain. The study of business models also addresses different forms of creativity, innovation, entrepreneurship and structure of organizations, funding systems, associations and relations with receivers.

Keywords Business model · Added value · News companies · Digital platforms

1 Introduction

The emergence of the Internet has transformed the process of news production for journalists (Mitchelstein and Boczkowski 2009; Deuze and Witschge 2018), consumption habits of audiences (Antunovic et al. 2018), and the flow, presence and reach of digital information. Journalists, audiences and news companies try to adapt their skills, preferences and organization structures to the new digital challenges that have impacted on their core activities (Aitamurto and Lewis 2013).

M. Goyanes (✉)
Universidad Carlos III de Madrid, Madrid, Spain
e-mail: mgoyanes@hum.uc3m.es

M. Rodríguez-Castro · F. Campos-Freire
Universidade de Santiago de Compostela, Santiago de Compostela, Spain
e-mail: m.rodriguez.castro@usc.es

F. Campos-Freire
e-mail: francisco.campos@usc.es

© Springer Nature Switzerland AG 2020
J. Vázquez-Herrero et al. (eds.), *Journalistic Metamorphosis*,
Studies in Big Data 70, https://doi.org/10.1007/978-3-030-36315-4_13

Online news, which are increasingly ubiquitous, atmospheric and market-oriented, have relatively lost their value, which affects directly to the payment intention of consumers (Goyanes et al. 2018), affecting the business models and income of most of online organizations (Myllylahti 2017; Cawley 2019). Within this context, there is a growing concern about audience´s perception on the value and reason-why of digital news. This study is intended to shed light on the perception of the value of digital news within a market-oriented context and their effects in the approach of new business models.

The article explores how audiences respond to the growing commodification and standardization of online news. By revising the situation of press, the audiovisual arena and digital platforms and their monetization models, we try to identify audience's attitudes towards digital news and the role of the Internet as a media outlet.

The book chapter is structured as follows: first, we make a review of the underpinning theory of economics of digital information, linking it with the growing need to charge for contents on the Internet, the payment intention of consumers and the recent research on paywalls. This helps us to understand the commodification and standardization process of digital information. Once the review is done, we shed light on the diversity of business models for digital platforms and the audiovisual arena, emphasizing the growing need to link their advances with the different forms of creativity, innovation and entrepreneurship of communication organizations around the world. The study contributes to the growing research on the reach and morphology of business models in the area of communication and its impact on consumption.

2 Standardization and Commodification of Digital Content and Its Impact on the Intention of Pay Per Contents

In the last decades, newspaper readers and advertising contents have been decreasing (Cawley 2019), while digital platforms have experienced a large increase. However, despite the digital take-off, online advertising and subscription income do not offset losses in printed press (Pickard and Williams 2014).

According to Edmonds (2017), US newspapers gained $1 in new digital advertising income for every $25 lost in printed advertising revenue. This positions digital and printed news companies in a difficult situation, which forces them, in many cases, to implement pay-per-content strategies based on business models that invite readers to pay (Bakker 2012). In fact, some suggest that paywalls are possible successful business models to monetize digital content, following the success of companies such as *The New York Times*, *Financial Times* and *The Wall Street Journal* (Benton 2018).

However, these media outlets seem to be more the exception than the rule. Besides, applying a successful business model to another media company is, at least, a risky operation. Most of news companies follow a trial-error based learning, adapting their organization structures and value proposals to the business and income model that best suit them (Arrese 2016).

Many companies had to cancel or postpone their paywalls and come back to free models, due to their inability to convince readers of the need to pay per contents. What lies behind this inability is, therefore, the creation of a valuable digital news' offer to put behind the wall. However, research on journalism suggests a standardization process of news (Picard 2009; Carlson 2015; Goyanes and Rodríguez-Castro 2018), especially on the Internet, which defies efforts of news companies to increase the money valuation of readers and their payment intention.

The goal of the study is to offer a better understanding of business models for digital platforms within a context of market-oriented and standard offer. The concept of market-oriented information or commodification refers to the process where news pass from being "products that satisfy individual and social needs to be products whose value is established by what they can bring to the market" (Mosco 2009: 132). On this basis, the news offer becomes generally a subsidized product–especially when it comes to online journals, as readers pay very little or nothing. At the same time, third parties, most of them advertisers, are the main income source, which means that the definition of quality is based more on "the most profitable popularity than something based on the less-shared professional standards" (McManus 1992: 790).

Therefore, it can be presumed than a valuable and high-quality news production, consumed by few readers, is ineffective for news companies. On the contrary, low-quality but highly consumed news are not. Commodification of online news has, therefore, a huge impact on the practice of journalism, which can be reflected, especially, on commercial pressures (Goyanes and Rodríguez-Castro 2018) and the report on social inequalities (McManus 1992).

This impact on commodification of news also permeates normative discussions about the main objectives of news in particular and journalism in general (Picard 2009). If a commodity is considered a good sold at a price in the profit-making market (Murdock 2011), then digital news could fit in well within this definition. Many editors, indeed, consider that their business is to sell commodities, that is to say, online news (Hantula 2015).

However, digital news also has a public function, reflected on the spirit of journalists to publish accurate and valuable information on public affairs and politics (Tuchman 1978). Therefore, despite the consensus on the growing commodification of the publishing business (Chen et al. 2015), journalism has to serve citizens, providing them with impartial, neutral and trustworthy information (Deuze 2005). Within this context, news companies are at the crossroads: on one side, they should be profitable to be on the business, relying on advertising as the main income source (McManus 1992), but they have at the same time the legitimate power to act as the fourth power, "driven by the search of trust" (Fisher 2014).

To attract readers' attention, news media use more and more sensational and exacerbated headers and contents, which intentionally creates fake and low-quality news (Chen et al. 2015). The commodification of online news therefore provides new technical and tactical developments to encourage interest and attention of readers when reading news, which are based on fraudulent strategies such as clickbait (Cable and Mottershead 2018).

In this regard, it could be said that clickbait is the most relevant result of the process of commodification of news, rooted in an organizational culture aimed at attracting readers' attention and encourage visitors to click in a link, regardless its quality and accuracy. Given that direct payments of readers have virtually disappeared and advertising income became crucial (Barthel 2015), news media publish more and more digital contents to generate dollars in advertising (Chen et al. 2015) at the expense of informing citizens.

A potential solution to limit news commodification, as we mention, is the implementation of pay-per-content strategies based on subscription models and micro-payments (Geidner and D'Arcy 2015). However, as demonstrated in previous research (Chyi 2005), digital news has been characterized by a demand curve at the price of zero. However, if the price increases even just a penny, the demand falls below zero. As there are a lot of free substitutes, the cross-price elasticity is high, so the free exchange envisaged by media outlets could result in a fall of the quantity, as the substitution effect would be activated (Chyi 2005). Only when price increases allow a sufficient subscription fee for the business development, pay-per-content strategies will be effective, which will limit the commercialization of news as a result of this process.

Also, the growing literature on standardization of news addresses mainly the effects that not-unique products and therefore potentially replaceable–digital news– could have on the demand when the price is greater than zero (Goyanes 2014; Chyi 2005; Picard 2009). Specifically, according to Picard (2009), both online and printed news are homogenous products and, therefore, not-unique. This means that, within a context of growing competence as the news business, digital news is fully and easily replaceable (Chyi 2005).

However, considering the successful value proposals of digital newspapers such as *The Wall Street Journal* and *The Financial Times*, there are some cases in which the digital offer is unique and, therefore, difficult to replace. Rather than the standardization of the media industry is, therefore, the standardization of a large part of digital news services. However, many media companies offer different value proposals difficult to imitate, which make their offer special, unique and different from competitors (Myllylahti 2017; Sjøvaag 2016).

As indirectly mentioned, the standardization of digital news is unavoidably linked to the perceptions of readers on the value of the news–economic but also informative. In this regard, journalists are ultimately responsible for turning facts into attractive, formal and informative news, based on the standards established in their field (Deuze 2005; Fisher 2014). Newsrooms play a key role in the design of a valuable news offer, capable of being captured later by media companies in the form of payment transactions of readers–paywalls.

Unfortunately, as pointed out by Picard (2009), the profession of journalism has become a commodity, and the professionalism of journalism and the teaching of journalism have determined the values and rules of the news, have commodified the product and have turn most of journalists into relatively interchangeable workers of a factory of information. The process, increasingly encouraged by the recent crisis in journalism and the instability of employment and professional conditions (Siles and Boczkowski 2012; Ekdale et al. 2015; Goyanes and Rodríguez-Gómez 2018), has determined the field for an increasingly similar news offer.

The consequences of standardization and commodification of digital news are essentially the loss of pecuniary value but also informative value (Myllylahti 2017). Recent market studies show that most readers are reluctant to pay for digital contents (Newman et al. 2016) and most of them also refuse future transaction. Specific factors such as age, income, previous payment, interest for news and consumption patterns are crucial to explain the phenomenon (an in-depth review may be found in Goyanes 2014; Fletcher and Nielsen 2017). Besides, the values of news such as exclusivity and authorship are crucial elements in the configuration of their pecuniary value (Goyanes et al. 2018).

Likewise, research on paywalls has shown the main strategies by media companies to monetize digital contents. For instance, Sjøvaag (2016) analyzed three digital Norwegian newspapers, showing that specialized content, such as local information, is usually paid, while syndicated content and immediacy news tend to be opened. Also, Myllylahti (2017) analyzed the main newspapers in Australia and identified the news and opinion articles as the main content in newspapers' paywalls.

Despite the existence of significant factors and values of news that, ultimately, result in the payment of readers, there is a general consensus that most of the literature on the intention of payment and paywalls actually investigates "the intention of not paying" (Goyanes et al. 2018: 10). Within this context, most of citizens are free consumers of digital news, as they take for granted their free access to information services and, therefore, reinforce the tendency to consider news as a public good and, as a result, develop a free culture on the Internet.

3 The Dominance of Digital Platforms: How to Monetize the Algorithm

The current informative context as a whole–press, radio, television and digital media– is experiencing a series of transformations resulting from the emergence of digital platforms. These platforms, among which are the GAFAN (Google, Amazon, Facebook, Apple and Netflix) or what van Dijck et al. (2018) call the Big Five (the same platforms, replacing Netflix for Microsoft), are reshaping the market of communication through dynamics of disruptive innovation (Christensen et al. 2015), which modify both business models of traditional media companies and the conception and development of contents and their consumption patterns.

The term digital platforms refers to their performance as a space of convergence between multiple agents (Miguel and Casado 2016), as will be developed below, and it is also the preferred denomination by platforms themselves. The choice is not neutral, but obeys the intention of these disruptive companies to establish the criteria from which to be judged and understood (Gillespie 2010), as well as regulated. Platforms had previously chosen to call themselves 'technology companies' instead of 'media companies' to avoid their publishing responsibilities (Napoli 2015). Also, due to their ability to connect other media and users, they are also called digital intermediaries.

Generally, the business model of digital platforms is based on an infrastructure where different agents converge; that is to say, they constitute multi-sided markets with different players such as users, content producers and advertisers, among others. Platforms expand bilateral and dominant markets which were so far of media companies, composed by audiences and advertisers, to include also agents such as advertising intermediaries and social institutions (Nieborg and Poell 2018).

The growth of these multi-sided markets depends on the development of effects, that is to say, according to Nick Srnicek, "the more users are using a platform, the more valuable that platform becomes for everyone" (2018: 46). This way, the bigger the number of registered users using Facebook, the more appealing will Mark Zuckerberg's social network be for advertisers willing to insert ads in the news feed, but also for media companies that want to expand their audiences through the dissemination of contents via social media. Digital platforms should take into account these effects of the network when designing their price structure through cross-subsidization strategies, in such a way that the parts generate business through the entrance of income (advertisers, for instance) to compensate those who access services free of charge, such as social media users.

These dynamics generate different business models within the platforms from the perspective of the final users, among which three basic variants stand out (Gabszewicz et al. 2015): offer of free contents and services, therefore, looking for alternative income; a model based on payment for access to contents (subscription, pay per individual product, etc.), which can be also complemented with other income sources; and a third model that combines the first two, the freemium, in which the access to a part of contents is free, but users have to pay to access the premium content.

Spotify represents this last model, as it offers free access to the whole catalogue, although with ads, and there is a premium version through subscription, without pauses, with a better audio quality, the possibility to listen offline and the removal of the shuffle-only mode. Meanwhile, Netflix has a business model based on a three-monthly plan subscription. All of them offer free access to the complete catalogue, with cheap rates and without commitment to remain, key elements for the popularization of the on-demand service (Izquierdo-Castillo 2015).

Despite the different articulations of the business models, all of them place the development and application of algorithms at the epicentre of their operations and success. When the user accesses platforms and surfs them, generates a series of information that platforms store and interpret through their algorithms, designed to give meaning to big data and allow their economic exploitation. Thus, users' data become a key element for the business model of platforms–that can use that information to customize ads–and for articulating the service and contents offered to users–through recommendation systems, prioritizing which contents will be more relevant for each user according to their interests.

The use of algorithms brings innumerable advantages to platforms, but their use also throws certain shadows. Bell and Owen, for instance, define the algorithm of the Facebook News Feed as "the single most controversial, influential and secretive algorithm in the world" (2017). As if it were the Coca-Cola formula, platforms keep the design and concrete functioning of their algorithms secret, in part because an increase in transparency around these formulas could negatively affect their business models and damage the dynamics of innovation (Napoli 2015).

However, the secrecy that permeates platforms' functioning, is a source of concern for publishers, who want to disseminate their contents through these digital inter-mediaries, due to the impossibility of designing strategies that guarantee the desired impact on these platforms. Digital intermediaries have become the new gatekeepers (Napoli 2015; Russell 2019), which results in the emergence of multiple challenges over how to relate with them, how to adapt contents to be usable, and how to reach a larger number of final users, and how to do it in such a way that their publishing principles are not altered.

Social networks, for instance, present a dilemma to editors. Although the last report from Reuters (Newman 2019) identifies a slight decrease in the consumption of news through social networks–due in part to changes in the Facebook algorithm, which prioritizes content of personal contacts at the expense of the exposure to publishing contents, 36% of respondents use Facebook to get informed.

The high number of users that can be accessed by digital platforms, especially if we take into account that many of them reach news contents incidentally while spending time in them for other reasons (Fletcher and Nielsen 2018), multiply the impact of contents, so that disseminating them through these intermediaries is too tempting. However, a successful presence of media companies in these platforms lies in adapting to their disruptive dynamics, based on engaging users to maximize the time they spend on the platform and, therefore, also the amount of data they generate. This results in clickbait strategies, in the production of potentially-viral contents and even the mass propagation of fake news (Braun and Eklund 2019).

The relationship between publishers and platforms, therefore, is problematic. In the context of platforms, and due to the commodification of the news–referred to in Sect. 2 of this chapter–each piece of news has value by itself and can individually circulate through different digital intermediaries. It breaks with the traditional vision of newsrooms and the prioritization of information according to editorial criteria (Nieborg and Poell 2018). This unbundling, therefore, is one of the most direct effects of digital intermediaries in the production and consumption of news contents, and forces news companies to reflect on the online circulation of published pieces (van Dijck et al. 2018).

Within the framework of unbundling and the relationship between publishers and platforms, the former can opt for two strategies in the dissemination of their contents (van Dijck et al. 2018). On the one hand, they can design networked strategies, based on the dissemination of links and headlines on different platforms with the aim of redirecting audiences to the media website, which allows the media to record the impact of each piece and monetize these visits.

On the other hand, there are native strategies, in which news content is offered directly on platforms, without the need to go somewhere else. This strategy implies that the media company, which is responsible for the editorial content, yields power over the management of the data registered by users and the monetization of the content.

These platforms, which initially tried to move away from this editorial perspective, have become more and more interested in disseminating news contents, especially if it is done in a native way. This way, different services oriented to reach dissemination agreements with other media companies emerged. In 2015, Facebook launched its function Instant Articles, which allows quick access to contents produced by third parties, with a visual and immersive format. These articles are accessible without quitting the platform and offer producers different alternatives for monetization. Other examples of tools linked to native strategies would be Discover from Snapchat, Google's Accelerated Mobile Pages, Apple News and the Live options on Facebook and Instagram.

The offer of this kind of functionalities opens up new possibilities to reach audiences, places publishers in a difficult situation, as they have fear of missing out something (Nielsen and Ganter 2018). Companies can opt for the adoption of native strategies for fear of getting list or being late to a good possibility of increasing their impact, though these strategies entail a lack of audience data, the migration of advertising revenues and even the dissolution of the publisher's brand (Bell and Owen 2017).

Although digital platforms occupy, judging by their economic results and their relationship with content producers, an omnipresent position, it is worth pointing out some of their limitations that contribute to nuance their hegemony. On the one hand, it can be glimpsed a certain saturation in the market of digital intermediaries. This phenomenon can be seen, for instance, in on-demand services.

One of the keys of Netflix's success was to allow access to its catalogue for a very economic rate, which resulted in an expansion throughout the world. However, it is unfeasible for a consumer to pay the subscription of all similar emerging services (Amazon Prime, HBO Go, Hulu, Apple TV, Movistar+, etc.), so there is a fragmentation of the market that leads to an increase in competition between platforms.

On the other hand, both this kind of on-demand services and the different social networks are facing challenges as regards regulation. Despite their attempts to evade publishing responsibilities, social networks such as Facebook, which are immersed in scandals such as the interferences and proliferation of fake news in the US presidential elections in 2016 (Bell and Owen 2017), will have to rethink their governance, aspiring to a balance between commercial interests and public interest (Napoli 2015).

As regards audiovisual digital platforms in the European Union, since the entry into force of the new Audiovisual Media Services Directive (European Parliament 2018), they are now forced to include in their catalogues at least 30% of European productions. The dominance of digital platforms, therefore, is starting to be regulated under the premises of public interest and cultural protection.

4 Journalism of Public Interest and Sustainability

The concept of business model includes a set of elements that, apart from the value proposal of the product or service, covers the needs and demands of customers or users, the conditions of the market in which it is developed, distribution channels, relationship system, resources and key associations, income sources and prices as well as structures and kind of costs (Osterwalder et al. 2005). The above-mentioned disruptive innovations in digital production and distribution have relativized or devalued the prioritization of the value proposal of the content in exchange for gratuity, data and sum of long-tail inputs (Anderson 2009).

The digital business model, which transformed the distribution channel and the commercialization based on the access to million customers, disrupted and damaged the model of traditional media against metamedia of the web 2.0 (Manovich 2005; Campos-Freire 2015). After that disruptive crisis, the reaction of traditional media can hardly be to do the same, but rather to reaffirm and reinforce the social value proposal of their contents. That is, beyond exchange and use values of the news activity, which are standardized in the commercialization of commodity models, there is the recovery of the social value of journalism and of quality entertainment contents. It is an indispensable value in journalism and in the media as a resource of public interest for the balance of the democratic and informative ecosystem of modern societies.

This social value, difficult to measure but essential for credibility and trust, as pointed out by Paul Steiger, creator of *ProPublica*, it is not just what the public wants, but the synthesis of the public interest and its utility for the public (Puentes-Rivera et al. 2018). It is a complex concept, but essential, for the relevance and recovery of news companies in the 21st century, as foreseen by Picard (2012). Some authors have also added nuances and adjectives to this journalism of public interest and social value (De Zúñiga and Hinsley 2013; Ferrucci 2015; Drok and Hermans 2016; Hermans and Glydensted 2019): good journalism, investigative journalism, constructive journalism, solutions journalism, commitment journalism, slow and contrastive journalism, etc.

The complexity in assessing the social value of the news makes their monetization and, therefore, the sustainability of the business model in the above-mentioned disruptive ecosystem. In the heart of that complexity lies the recovery of trust, interest and the overcoming of that tiredness of news registered by surveys and reports on the era of abundance of information (Palmer and Toff 2018).

In the search for a remedy to the crisis of traditional media, which took with it thousands of printed publications in western countries, companies and many forced entrepreneurs coming from news staff and then fired, started to propose alternatives to find solutions for a sustainable journalism. Also, public debates were opened in parliamentary and government institutions to seek solutions to the crisis of journalism, inasmuch as it is a source of mediation and legitimacy of modern democracies, especially as regards a new expanded concern about the virality of fake news in digital networks and the Internet.

From 2017, France, Germany, Italy and the UK have pioneered regulation on fake news in Europe. After two years of debate at the European Parliament, the European Union approved in November 2018 the Directive 1808 on the provision Audiovisual Services, which regulates platforms against traditional broadcasters. In 2019, another Directive on intellectual property was agreed to protect creators' and journalists' copyright as regards digital networks and Internet infomediaries. The new Creative Europe program for 2021–2027 includes into their projections to support the press and the traditional media sector as relevant activities of cultural and creative industries.

In different states, France plays a pioneering role in support policies to the press, based on an integral industrial concept, which covers its entire value chain, from publishing, printing, distribution and delivery of publications to readers. In 2008, Nicholas Sarkozy opened the institutional debate on the 'general states of the press', in charge of people from society and the industry, to present a report of measures to the government and the Senate and thus regulate public aids to the sector, which amounted to 284 million in 2009. Subsidies to the press have been maintained with different lines, reports and assessment of the most critical aspects. The last one was made by Schwartz and Terraillot (2018) for the Ministry of Economy and Finance on the distribution system of the French press, which proposes ten reform and support measures.

Also, the UK government, previous commitment of the Prime Minister Theresa May before the main publishers of the country, commissioned a study on the future of journalism and the sector to Professor Francess Cairncross (2019), who concluded the report with nine support proposals to the sustainability of the sector.

Other countries maintain and promote institutional policies for the sustainability of the press and the future of journalism. These include the Netherlands, Denmark, Norway, the Flemish region of Belgium, Canada and Australia. In the last two countries, reports, proposals for measures and debates were made in their respective parliamentary chambers, highlighting the need to support local information, their newspapers and journalism as a service of general interest (Vine 2017). Some European countries have begun to apply digital taxes to the platforms to contribute with these funds to the sustainability of precarious means of information of general interest.

In the United States, the support to the press and newspapers, which suffered a massive hemorrhage of headings and job destruction, is channeled through foundations and local institutions. A quarter of the respondents surveyed in the trend study by Newman (2019) from the Reuters Institute consider that public or institutional support is necessary to maintain quality journalism; 29% of them believe that they could come from foundations and non-profit organizations, 18% expect them to be from the contribution of digital platforms and 11% say that from governments.

5 Conclusions

Social value and innovation, in the form of knowledge and creative intelligence, are the heart of the journalism business model. Public service media organizations, following the strategy of its main state and regional or regional models of European states, try to revive and combine their original principles of the triad–to inform, to educate and to entertain–by John Reith when the *BBC* emerged, with the six core values established by the EBU in 2014–universality, quality, independence, diversity, accountability and innovation–(EBU 2014) and the adaptation to other new emerging values of the new society. Journalism and quality information are an essential requirement for public service media.

The sustainability of journalism and quality information, in addition to the protection of the traditional dual system of generating resources from the sale of contents and advertising, also requires institutional support as well as adaptation to the model of metaservices in which they fit, provided that contribute and not deteriorate their value, the various forms of income–product, digital and mobile content, premium, open access, programmatic and native advertising, sponsorship, bartering, membership, crowdfunding, foundations, public aid, events, augmented reality, artificial intelligence, gamification, Internet of things, blockchain, etc.

Acknowledgements The chapter belongs to the activities of the research project RTI2018-096065-B-I00, from the Spanish State Program R + D oriented to the challenges of society from the Ministry of Science, Innovation and Universities (MCIU), Agencia Estatal de Investigación (AEI) and the European Regional Development Fund (ERDF) on *New values, governance, funding and public media services for the Internet society: European and Spanish contrasts*.

References

Aitamurto T, Lewis SC (2013) Open innovation in digital journalism: examining the impact of open APIs at four news organizations. New Media Soc 15(2):314–331

Anderson Ch (2009) La economía de la larga cola. Empresa Activa, Barcelona

Antunovic D, Parsons P, Cooke TR (2018) 'Checking'and googling: stages of news consumption among young adults. Journalism 19(5):632–648

Arrese Á (2016) From gratis to paywalls: a brief history of a retro-innovation in the press's business. Journalism Stud 17(8):1051–1067

Bakker P (2012) Aggregation, content farms and Huffinization: The rise of low-pay and no-pay journalism. Journalism Pract 6(5–6):627–637

Barthel M (2015) Newspapers: fact sheet. Retrieved from http://www.journalism.org/2015/04/29/newspapers-factsheet

Bell E, Owen T (2017) The platform press: how Silicon Valley reengineered journalism. C Journalism Rev. Retrieved from https://www.cjr.org/tow_center_reports/platform-press-how-silicon-valley-reengineered-journalism.php

Benton J (2018) The New York times is on pace to earn more than $600 million in digital this year, halfway to its ambitious goal. Retrieved from http://www.niemanlab.org/2018/11/the-new-york-times-is-on-pace-to-earn-more-than-600-million-in-digital-this-year-halfway-to-its-ambitious-goal/

Braun JA, Eklund JL (2019) Fake news, real money: ad tech platforms, profit-driven hoaxes, and the business of journalism. Digit Journalism 7(1):1–21. https://doi.org/10.1080/21670811.2018.1556314

Cable J, Mottershead G (2018) Can I click it? yes you can: sport journalism, twitter, and clickbait. Ethical Space Int J Commun Ethics 15(1/2):69–80

Cairncross F (2019) A sustainable future for journalism. Retrieved from https://www.gov.uk/government/publications/the-cairncross-review-a-sustainable-future-for-journalism

Campos-Freire F (2015) Adaptación de los medios tradicionales a la innovación de los metamedios. El Profesional de la Información 24(4):441–450

Carlson M (2015) The robotic reporter: automated journalism and the redefinition of labor, compositional forms, and journalistic authority. Digit Journalism 3(3):416–431

Cawley A (2019) Digital transitions: the evolving corporate frameworks of legacy newspaper publishers. Journalism Stud 20(7):1028–1049

Chen Y, Conroy NJ, Rubin VL (2015) Misleading online content: recognizing clickbait as false news. In: Proceedings of the 2015 ACM on workshop on multimodal deception detection, ACM, 15–19

Christensen CM, Raynor ME, McDonald R (2015) What is disruptive innovation. Harv Bus Rev 93(12):44–53

Chyi HI (2005) Willingness to pay for online news: an empirical study on the viability of the subscription model. J Media Econ 18(2):131–142

De Zúñiga H, Hinsley A (2013) The press versus the public. What is "good journalism"? Journalism Stud 14(6):926–942

Deuze M (2005) What is journalism? professional identity and ideology of journalists reconsidered. Journalism 6(4):442–464

Deuze M, Witschge T (2018) Beyond journalism: theorizing the transformation of journalism. Journalism 19(2):165–181

Drok N, Hermans L (2016) Is there a future for slow journalism? the perspective of younger users. Journalism Pract 10(4):539–554

EBU (2014) Public service values, editorial principles and guidelines

Edmonds R (2017) Newspapers get $1 in new digital ad revenue for every $25 in print ad revenue lost. Retrieved from https://www.poynter.org/reporting-editing/2012/newspapers-print-ad-losses-are-larger-than-digital-ad-gains-by-a-ratio-of-25-to-1/

Ekdale B, Tully M, Harmsen S et al (2015) Newswork within a culture of job insecurity: Producing news amidst organizational and industry uncertainty. Journalism Pract 9(3):383–398

European Parliament (2018) Directive (EU) 2018/1808 of the European parliament and of the council of 14 November 2018 amending directive 2010/13/EU on the coordination of certain provisions laid down by law, regulation or administrative action in Member States concerning the provision of audiovisual media services (Audiovisual Media Services Directive) in view of changing market realities

Ferrucci P (2015) Public journalism no more: the digitally native news non-profit and public service journalism. Journalism 6(7):904–919

Fisher M (2014) Who cares if it's true?. Retrieved from http://www.cjr.org/cover_story/who_cares_if_its_true.php?page=all

Fletcher R, Nielsen RK (2017) Paying for online news: a comparative analysis of six countries. Digit Journalism 5(9):1173–1191

Fletcher R, Nielsen RK (2018) Are people incidentally exposed to news on social media? a comparative analysis. New Media Soc 20(7):2450–2468

Gabszewicz JJ, Resende J, Sonnac N (2015) Media as multi-sided platforms. In: Picard RG, Wildman SS (eds) Handbook on the economics of the media. Edward Elgar Publishing, Cheltenham and Northampton, pp 3–35

Geidner N, D'Arcy D (2015) The effects of micropayments on online news story selection and engagement. New Media Soc 17(4):611–628

Gillespie T (2010) The politics of "platforms". New Media Soc 12(3):347–364

Goyanes M (2014) An empirical study of factors that influence the willingness to pay for online news. Journalism Pract 8(6):742–757

Goyanes M, Artero JP, Zapata L (2018) The effects of news authorship, exclusiveness and media type in readers' paying intent for online news: an experimental study. Journalism (forthcoming)

Goyanes M, Rodríguez-Castro M (2018) Commercial pressures in Spanish newsrooms: between love, struggle and resistance. Journalism Stud 20(8):1088–1109

Goyanes M, Rodríguez-Gómez EF (2018) Presentism in the newsroom: how uncertainty redefines journalists' career expectations. Journalism (forthcoming)

Hantula K (2015) Four + one truths about paywalls. Retrieved from http://livinginformation.fi/en/articles/four-plus-one-truths-about-paywalls

Hermans L, Gyldensted C (2019) Elements of constructive journalism: characteristics, practical application and audience valuation. Journalism 20(4):535–551

Izquierdo-Castillo J (2015) El nuevo negocio mediático liderado por Netflix: estudio del modelo y proyección en el mercado español. El Profesional de la Información 24(6):819–826

Manovich L (2005) El lenguaje de los nuevos medios de comunicación: la imagen en la era digital. Paidós, Barcelona

McManus J (1992) What kind of commodity is news? Commun Res 19(6):787–805

McNair B (2000) Journalism and democracy. An evaluation of the political public sphere. Routledge, London

Mitchelstein E, Boczkowski PJ (2009) Between tradition and change: a review of recent research on online news production. Journalism 10(5):562–586

Miguel JC, Casado MÁ (2016) GAFAnomy (Google, Amazon, Facebook and Apple): the big four and the b-ecosystem. Dynamics of big internet industry groups and future trends. Springer, Cham, pp 127–148

Mosco V (2009) The political economy of communication. Sage, London

Murdock G (2011) Political economies as moral economies: commodities, gifts, and public goods. The handbook of political economy of communications. Wiley-Blackwell, Malden, pp 11–40

Myllylahti M (2017) What content is worth locking behind a paywall? digital news commodification in leading Australasian financial newspapers. Digit Journalism 5(4):460–471

Napoli PM (2015) Social media and the public interest: governance of news platforms in the realm of individual and algorithmic gatekeepers. Telecommun Policy 39(9):751–760

Newman N, Fletcher R, Levy DAL et al (2016) Reuters institute digital news report 2016. University of Oxford, Oxford

Newman N (2019) Journalism, media, and technology trends and predictions 2019, Reuters Institute for the study of journalism. University of Oxford, Oxford

Nieborg DB, Poell T (2018) The platformization of cultural production: theorizing the contingent cultural commodity. New Media Soc 20(11):4275–4292

Nielsen R, Ganter SA (2018) Dealing with digital intermediaries: a case study of the relations between publishers and platforms. New Media Soc 20(4):1600–1617

Osterwalder A, Pigneur Y, Tucci C (2005) Clarifying business models: origins, present, and future of the concept. Commun Assoc Inf Syst 16

Palmer R, Toff B (2018) From news fatigue to news avoidance. NiemanLab Predictions for Journalism 2019. Retrieved from https://www.niemanlab.org/2018/12/from-news-fatigue-to-news-avoidance/?relatedstory

Picard RG (2009) Why journalists deserve low pay. Christ Sci Monit 19(9):1–6

Picard RG (2012) La creación de valor y el futuro de las empresas informativas. Por qué y cómo el periodismo debe cambiar para seguir siendo relevante en el siglo XXI. Media XXI, Porto

Pickard V, Williams AT (2014) Salvation or folly? the promises and perils of digital paywalls. Digit Journalism 2(2):195–213

Puentes-Rivera I, Campos-Freire F, López-García X (2018) Periodismo con futuro. Media XXI, Lisbon

Russell FM (2019) The new gatekeepers: an Institutional-level view of Silicon Valley and the disruption of journalism. Journalism Stud 20(5):631–648

Schwartz M, Terraillot F (2018) Dix propositions pour moderniser la distribution de la presse. Rapport au ministre de l'Économie et des Finances et à la ministre de la Culture. Retrieved from https://www.ladocumentationfrancaise.fr/var/storage/rapports-publics/184000497.pdf

Siles I, Boczkowski PJ (2012) Making sense of the newspaper crisis: a critical assessment of existing research and an agenda for future work. New Media Soc 14(8):1375–1394

Sjøvaag H (2016) Introducing the paywall: a case study of content changes in three online newspapers. Journalism Pract 10(3):304–322

Srnicek N (2018) Capitalismo de plataformas. Caja Negra Editora, Buenos Aires

Tuchman G (1978) Making News—a study in the construction of reality. The Free Press, New York

van Dijck J, Poell T, de Waal M (2018) The platform society: public values in a connective world. Oxford University Press, New York

Vine P (2017) When is a journalist not a journalist?: negotiating a new form of advocacy journalism within the environmental movement. Pac Journalism Rev 23(1):43–54

Manuel Goyanes Assistant Professor in the Department of Media and Communication at Carlos III University (UC3 M) in Madrid. Dr Goyanes' main research focus is in media management and sociology of communication sciences. He was visiting fellow in the Department of Media and Communications at the London School of Economics and Political Science. His work has appeared in journals such as *International Journal of Communication, Journalism Practice* or *International Journal on Media Management*.

Marta Rodríguez-Castro Ph.D. Candidate and predoctoral researcher at Universidade de Santiago de Compostela. In her doctoral thesis she is studying Public Value Tests in Europe in order to establish a set of quality indicators and to assess and discuss its possible introduction in Spain. She was visiting fellow at the Center for Media, Data and Society in the Central European University (Budapest, 2018).

Francisco Campos-Freire Full professor of Journalism in the Faculty of Communication Sciences of Universidade de Santiago de Compostela. His main research lines are the management of media companies, the study of the impact of social networks and innovation on traditional media, and the funding, governance, accountability and transformation of European Public Service Media. In the professional field, he has worked as editor and director in several Spanish newspapers and television and radio companies.

New Forms of Journalistic Legitimization in the Digital World

Laura Solito and Carlo Sorrentino

Abstract The 'digital revolution' is transforming the journalistic field. From sequential processes—selection, verification, hierarchization, presentation, fruition—journalism is moving on to a 'despacialized simultaneity' in which production, distribution and consumption are intertwined. These evolutions are modifying the legitimation of the credibility and authority of journalism, which historically resided in its capacity to guarantee a public service whose hierarchical nature was ensured by exclusivity. The operational procedures through which journalism built its reputation must be rethought to include citizens more. Participation and transparency become central principles, despite the lack of clarity as to which new practices will substantiate them.

Keywords Journalistic legitimization · Citizenship · Participation · Transparency

1 Introduction

Digital communication expands the social circles frequented by single individuals, it enables and facilitates the diversification of social roles, for decades a feature of modern times, and defines the individualization process. These evolutions reflect significantly on the representation of reality produced by journalism, but even further on the citizens' relationship with journalism as an institution.

In this chapter, we will analyse these developments in order to highlight how the principles legitimizing journalism are changing too, as it is less and less called upon to 'transport' contents from the sources producing them to the public of users (Peters and Broesma 2013), and is instead involved in a complex process of negotiation

While devised jointly by the two authors, the first and third sections were written by Laura Solito and the introduction and second section by Carlo Sorrentino.

L. Solito · C. Sorrentino (✉)
Dipartimento di Scienze Politiche e Sociali, Università degli Studi di Firenze, Florence, Italy
e-mail: carlo.sorrentino@unifi.it

L. Solito
e-mail: laura.solito@unifi.it

© Springer Nature Switzerland AG 2020
J. Vázquez-Herrero et al. (eds.), *Journalistic Metamorphosis*,
Studies in Big Data 70, https://doi.org/10.1007/978-3-030-36315-4_14

with sources and publics increasingly able to take part in the communication game. Indeed, also thanks to the lowering of the entry barriers caused by the so-called disintermediation process, both the sources and the public manage their presence in the public space through distinct, more or less effective communication codes which are nevertheless able to expand what is considered newsworthy, that is, the topics and social subjects dealt with by journalism (Benson and Neveu 2004; Sorrentino 2006). The phrase used by McNair (2006, 2018) to describe the progressive fragmentation and destructuring of the journalistic field (Mancini 2013) is "cultural chaos", a chaos resulting from the fact that each one of us continuously receives information from:

(1) traditional mainstream newspapers;
(2) sources increasingly professionally equipped to manage their own communication needs themselves, through the range of channels now available: from web sites to social networks;
(3) other components of the immense plethora of users, now able to remediate (Bolter and Grusin 1999) messages to their circuits of friends and followers using their own social walls.

The journalistic field is crammed with voices, of varying degrees of persuasion and attraction, which can inhabit the mediatized public sphere (Thompson 1995), potentially democratizing it, thanks to the possibility of enriching the public discourse. At the same time, however, one of the main characteristics of journalism is fading away: the broad consensus as to the principles for selecting which topics are to be spoken of, a consensus enabled by the clear evidence of what is relevant and of interest to the public. Indeed, with a more clearly defined and outlined hierarchy of values, it becomes easier to identify the concept of public interest establishing the importance of the events to be covered by journalism (McQuail 2013), and therefore it is simpler to activate consensus on the basis of a more distinct universe of values (Deuze 2015). The primary definers, as Hall et al. (1978) saw it, the political and economic elites able to impose topics and priorities on the agenda, are few and well legitimized by the central social role that they are clearly acknowledged.

In recent decades, the mass individualization process has progressively eroded these clear-cut hierarchies. What has to be made public has become less evident and the relevance with which this publicity takes places more faded. Digital communication further accentuates this process through the multiplication of the number of both broadcasters and recipients. 'The universe of the tacit presuppositions' upon which journalistic stories rest is being constantly bled dry (Benson and Neveu 2004). The abundance of journalistic depictions is fragmented, the public segmented and broken up. The continual rotation of topics highlights the selective nature of journalism. Its function of reconstructing reality makes it appear more biased, highlighting the limits of the presupposition historically at the basis of journalistic legitimation: objectivity of the facts. A central role is assumed by the negotiating process, based on a dense web of relations with sources increasingly deft at managing the media logics and with a public progressively less willing to completely rely on the definitions of situations provided by the press, now stripped of their certainty by the multiple truths given by the range of information on offer (Lorusso 2018). Not only this, owing to the disintermediation favoured by digital means, the users are now transformed into 'prosumers': consumers and producers at the same time.

Returning to a metaphor often used to describe the institution of journalism, while before newspapers were the streets where journalists exposed topics and subjects according to a stable and easily identifiable hierarchy, now, in the digital space, all actors take to the streets, without waiting for someone to do so in their stead. Hence, journalism is called upon to rethink its mission: more than selecting what to lay bare in the street, it has to create order in the crowd of actors, events and phenomena which are already in the street, crowding out this space, each one claiming visibility and centrality. However, it has to do so by negotiating and dialoguing with these interlocutors according to different logics from the past, in line with the new sphere of communication drawn by digitalization.

We will analyze the transformation in the relations between citizens and journalism while first of all dwelling on the evolutions in the concept of citizenship; then we will look into the consequences of these evolutions on journalism. Lastly, we will round off our reflection by setting out the principles of legitimation which we deem the necessary foundations for a new information pact between journalism, as the institution charged with recounting reality, and the citizens who make use of it.

2   The Demanding Citizen

Citizens have growing demands for information because knowledge is becoming an inescapable resource for them to move consciously within complex societies. This wealth of information is at the basis of the definition of a plural identity whose characteristics are reflected on the public and political dimension, redefining the concept of citizenship, which is becoming multidimensional. Reflections on the formal and institutional aspects of citizenship go hand in hand with culturalist and relational aspects.

In his classic study on citizenship, Marshall (1950) distinguishes three foundations characterizing citizenship in democratic societies: (1) civil rights, described as necessary for individual freedom—freedom of speech, religion, thought, right to ownership and to obtain justice–; (2) political rights, consisting of the possibility of having active or passive access to exercising political power; (3) social rights, ensuring economic security and well-being, also through the acquisition of a level of formal education or use of certain social services. The three foundations form dimensions that can expand and retract depending not just on the quantity of rights acquired, but also their quality. This variety shows how the concept of citizenship is complex and slippery; hence, it is a good idea to stick to an 'extensive' conception that adds cultural and 'technological' elements to the formal requirements described by Marshall (Rodotà 2013).

What substantiates cultural citizenship, first of all, is an affective dimension, defined by collective groups of belonging in which every citizen participates. Citizenship takes on plural forms and contains relational elements built in the various spaces of individuals' lives which—as already said—progressively expand from the family, to the neighbourhood, to the workplace, to the different forms of socialization and consumption of free time, and finally to the media spaces, more recently boosted by the multiplication guaranteed by digital communication.

Citizenship is increasingly clearly becoming something that is created day by day, rather something that is determined by formally and definitively acquired rights. Dahlgren (2009) quite rightly speaks of *achieved citizenship* rather than *received citizenship*. Immersed in society, every individual acquires stimuli, practices and resources which constitute his or her status as a citizen.

Dahlgren speaks of *civic agency* as the distinct way of building citizenship, played out within a civil society meant as a distinct space from the private sphere, but also from the political and economic institutions. It is a new relational context which readapts Habermas' concept of public sphere in light of the transformations produced by the gathering of many new subjects and social phenomena in a new public space, where individuals construct their relationality.

Journalism, and the media more in general, play an important role in this process of redefining the public space. On one hand, the use of journalistic products is included among the practices characterizing the democratic and civil development of a social context (Putnam 2000; Cartocci 2007), and it is positively correlated with participation in public life, as well as with the construction of broad and intense social relations. On the other hand, the same expansion of the forms of action made possible by a greater capacity to compare cultures, lifestyles and social practices makes experiences more superficial, limiting cultural consensus and making social cohesion and the feeling of belonging more fragile.

Political and cultural forms of belonging come apart and reform on the basis of much more complex relational networks. Individuals draw their values, which give rise to less solid forms of belonging these days, as well as the attitudes and behaviours which direct their lifestyles, from more and more groups. Citizenship is practised through many activities, which might be apparently distinct or distant from those usually counted in so-called conventional participation.

The increasingly complex requirements of citizenship produce an apparently contradictory effect: the increase in interest in democratic life, albeit to different extents from one citizen to the next, corresponds to a growing dissatisfaction and disillusionment with the effectiveness of the political responses made, which leads to mistrust and the 'critical citizen syndrome' (Norris, 1999) both concerning the political system and the media. What is more, this mistrust is augmented by the cynicism with which the media represent reality (Cappella and Jamieson 1997).

The reflections of Rosanvallon (2006) lie at the juncture between the possible outcomes of this cynicism. For the French scholar, the attention with which watchful and sceptical citizens survey and judge institutional power can be a stimulus so that the 'democracy of expression', in which many can take the floor and make claims, transforms into a 'democracy of involvement' meant as the active identification of new practices of intervention which enable expressivity to become a 'voice' (Couldry 2010), which Rosanvallon calls positive distrust; or it can degenerate into diffidence, in which the objective becomes to limit power with the consequent paralysis of the political field.

Journalism is a central institution to practise this watchfulness. Schudson effectively grasps this element when he theorizes monitorial citizens,

> perhaps better informed than citizens of the past ... but there is no assurance that they know at all what to do with what they know ... watchful, even while he or she is doing something else ... citizenship now is a year-round and day-long activity as it was only rarely in the past (Schudson 1996: 451).

This more superficial but continual attention to the surrounding reality enables the expression of a more conscious and refined citizenship, in which the distribution of people's lives in different social spheres is due to the growth of individual freedom and the consequent attention to safeguarding individual rights.

Schudson is well aware of how these evolutions can lead to a progressive reduction in the role of the parties and the contemporary rise of the media, with their tendency to personalize politics and emphasize conflict. Nevertheless, he deems that these transformations promote a mass democracy in which single people progressively become familiar with individual rights, the rights fought for by the new social subjectivities embodied by the movements that came about in the 1960s, rights which do not weaken community belongings but redefine their forms.

For Schudson this new form of citizenship—based principally on individualization—requires consciousness on the part of every person as to other people's rights. Hence, a sensitivity needs to be developed towards the interdependence between their actions and those of the people with whom they enter relations. We are not far from the 'critical citizen', who, as Pippa Norris has it, adapts his or her belonging to the particular community into different forms and various levels, showing a sense of belonging both to the nation and to the democratic ideals that it represents. For Norris too, dissatisfaction highlights interest and expresses an assessment resulting from attributed relevance, rather than apathy and disinterest.

Watchfulness, control, monitoring and assessment are all dimensions that lessen the deference towards the authorities and institutions; but above all they make citizens more demanding and more exposed to disappointment.

All in all, the progressively multidimensional nature of citizenship must not be measured in terms of the quantity of participation, to then complain of its scarcity, but of a new consistency; we should seek to observe and understand the many and different ways in which it can be expressed in a society characterized by multiple, albeit less solid, belongings which embody the mass individualization.

3   The News: From Product to Process

While the dimensions of citizenship and the ways in which it defines itself are changing, it also appears inevitable to rethink the role of journalism and, above all, the relationship between citizens and information. Tellingly, Peters and Broesma (2013) underline how rethinking the function of journalism means rethinking citizenship.

The greater dynamism of a citizenry less and less founded on definitive belongings, and no longer only played out in the traditional places of physical and cultural proximity (family, neighbourhood, school), produces an extension of the forms in which the public discussion is structured (Riegert 2007). Creating discourse, the *talkative society* (Dahlgren 2009) which produces that which Bakardjieva (2010) defines as *subactivism*—civic participation made of negotiations, disputes and agreements around which there are the right rules to order social life, and establish the ethical and moral perspectives to pursue–, acquires importance. It is only later on that these discussions take on a political dimension and become political behaviour and activity.

In creating discourse, journalism is one of the main institutions through which to put one's identity at stake, build one's own lifestyles, outline specific cultural perspectives. Information becomes the setting within which citizens live, an ongoing, background activity. It loses a large part of its traditional characteristic of being a deliberate act—going to buy the newspaper, tuning into the evening news—to transform into an ecosystem where the constant practice of updating information is now a given fact.

But precisely the progressive 'naturalization' of the information process, almost taken for granted and spontaneously encompassed in our everyday practices, dulls the awareness of the necessity to rethink the way in which the sources, journalistic system and public interact, both at the professional level *and* in the social representations of public opinion, as this process is still centered around the line running from the source to the public through journalistic mediation. Instead, now every component of the public sphere is part of a flow which continually alternates information coming directly from the sources (suffice it to think of the frequency with which political actors use the social networks to communicate with their voters) with that mediated by the journalistic system, and other information still that arrives directly from friends and followers who post additional stories, comments, reports and articles taken from other media on their social walls. So, more and more often, a significant portion of journalistic mediation takes place outside the journalistic organizations and outside the consolidated forms. Thus, this forms a very varied and differentiated range of practices, as shown by the personal accounts through which sources, journalists and citizens interact (Anderson et al. 2012; Deuze and Witschge 2018).

Singer (2018) sums up the evolutions caused to journalism by digital communication in 5 'I's: immersive, interconnected, individualized, iterative and instantaneous.

It is more and more difficult to reconstruct the way we found out about a piece of news: from the radio, TV, through Facebook, in the newspaper? We are *immersed* in information, we are *immersed* in the media; we really are merging with them (Deuze 2012). It is now obsolete to think of messages moving from issuers to receivers thanks to journalists who direct the traffic. An *interconnection* is produced that makes each of us a more or less bright dot in potential contact with every other link in the network, even though in reality we actually only interact with a limited and repetitive number of them. In these interactions, the distinction between producer and consumer also becomes more blurred; not only because each of us—above all through the social networks—produces information, but also because of the frequency with which we retransmit media mainstream information that we have just consumed to others. The wealth of information in circulation results in much more personalized multimedia diets. Even though the forms of consumption overlap, each of us assembles the set of information received in a totally *individualized* way. Furthermore, the news is constantly updated, and linked to other things. Hence, it is an *iterative* form of communication. While classic information is clearly outlined in space and time, with fresh news continually replacing previous stories, digital information is accumulative, it puts together information that is always easy to retrieve. But, above all, digital journalism is *instantaneous*, immediate.

By going beyond a linear vision of the information process, the communication transmission model based on the central position of the issuer, who vertically 'fills' the receiver with contents, is undermined. Carey (1989)—as long as 30 years ago—preferred the paradigm of the ritual, or sharing, according to which *communication is a symbolic process where reality is produced, consolidated, corrected and transformed*. This conception brings out the role of the public sphere as the collective place where civil society and State come into contact, the place of negotiation and where identities are compared (Pizzorno 2007). Following this paradigm, the news is seen less and less as a product and more and more as a process (Robinson 2009), defined by intense and changing relations with the sources and the public. The digital environment brings out this tendency precisely because the sources, the public and the journalists are inextricably interlinked, thanks to the many directions and the immersive nature of the flows. Everyone takes part in a perennially ongoing communication game (Hjarvard 2013). The great quantity of news generated, and above all the forms and ways in which it is divulged make for a less clear distinction between the function of providing citizens with information so that they can act and take part in political life, and the function of formulating the knowledge of a public of consumers in order to entertain them, aiming more at their emotional involvement (Hanitzsch and Vos 2018).

Moreover, all one has to do is speak to any journalist about his or her day-to-day activities to hear how, now, a journalist's work is not finished when pressing send to publish an article, as was the case in the past. Now, instead, it is just beginning, owing to the possibility/need to follow the piece through the myriad channels that it will follow, in the variety of appropriate languages that each of these channels requires.

4   How to Legitimate a New Information Pact

Such a far-reaching redefinition of practices is reflected on the relations between the various social actors in the negotiation. As many studies on *participatory news* have been claiming for some time (Deuze et al. 2007; Heinrich 2011), the crux lies in the progressive modification of the elements making up the relationship between the institution of journalism and the citizens. But if the nature of the pact changes, the presuppositions upon which this pact is defined have to change too. In other words, there is a change in the basis legitimating journalism as an authoritative institution identifying the events, topics and subjects towards which the public opinion's attention is to be directed.

Traditionally, journalistic legitimation has been twofold. First of all, the journalist guaranteed the truth of what was being claimed through verification activities (Tong 2018), one of the central elements of journalistic work. Think of the great deal of above all Anglo-Saxon literature identifying which operating procedures can elevate the journalist's work, such as the use of objectivity and impartiality. Tellingly defined as strategic rituals (Tuchman 1973; Schudson 1978), since they are followed by information professionals precisely to guarantee their lack of ties to their interlocutors: the sources but above all the public. As Kovach and Rosenstiel (2001) efficiently summed up: *the method, not the journalist, is objective*. These rituals were formalized with the precise aim of safeguarding the journalists' freedom of action, but also the ways in which this freedom was guaranteed. It was an objectivization of everyday professional practices necessary to guarantee the credibility of journalism.

In addition, journalism has always been acknowledged as having another distinct aspect: that is, to rank or define the level of significance of an event, a topic or a particular social actor. By establishing priorities, journalism also defines the degree of public interest in every piece of news. Of course, as mediological research has recognized for decades now, the journalistic system negotiates with many other social actors when carrying out this action. Nevertheless, recognizable and recognized standards have been established over time. Indeed, the concept of mediatization derives precisely from the growing relevance attributed to the operating modes used by the journalistic institution.

Verifying and ranking were two significant sides of journalistic legitimation. Thanks to the identification of procedures which have become consolidated over time, they came—undisputedly—to embody credibility standards for the verification activities, as well as establishing—through the quantity of space and time attributed to a piece of news—what the public opinion should know. But we also need to consider the way in which these activities were carried out. According to Carlson (2017), at length the legitimation of the credibility and authority of journalism resided in the capacity to guarantee a public service whose hierarchical nature was ensured by exclusivity. It was necessary to build a third place between the sources and users which could define and certify which events were of public interest (McQuail 2013). The journalistic procedures responded to a widespread need to recognize institutionalized practices for confirming the truth and significance of a news story.

Hence, the validity of these operating procedures derived from their necessity. But it was this same process that favoured the transformation of journalism into something more than the belt conveying news from a source to the public. Indeed, deciding what to discharge onto this conveyor belt or what to transport on it more frequently and quickly, gave journalism a central function in defining the priorities and highlighting which topics to put to the discussion of public opinion.

This function worked so long as the other points in the negotiation had few possibilities of directly coming into contact with each other and so journalism was able to preserve that hierarchical priority mentioned earlier. But the digital revolution made the boundaries between the roles of the various actors in the journalistic negotiation more movable. Suffice it to think how the growing distribution of news through search engines and social networks is completely redefining journalism's central task of 'shaping information', that is, assembling the entire body of information worthy of making the news. Now, instead, single pieces of news emerge one by one, minute by minute, all connected together like the pieces of a jigsaw puzzle. Thus, journalism is progressively losing its monopoly on the production and divulgation of news, which formed the basis of its authority, and acceptance of the legitimacy of its *modus operandi* (Tong 2018).

But, in this way, the distinction between information and communication—based on two significant presuppositions: the centrality of the facts and their truthfulness—is becoming blurred.

In journalism (in the same way as in common language), we refer to facts as given entities. The Latin etymology of the word, however, points to something that has been made, from the verb *facere*, to make, underlining the processual nature of building the action present in every fact (Knorr Cetina 1981). It is indeed the journalist's job to verify the givenness of the 'fact', which translates into checking the doubtlessness of that 'fact'. Journalistic verification not only attributes the label that something 'really happened', but also that it is relevant and of public interest based on a solid agreement as to the meaning to attribute to the 'fact', that is, the degree of stabilization achieved at the social level from the interpretations built up around it (Lorusso 2018). Infant mortality is an 'intolerable fact' in our societies where it has been progressively reduced until its near disappearance. Instead, it still appears a 'natural fact' in situations where it is an everyday occurrence. Consequently, the death of a child has a higher degree of newsworthiness depending on the context where it happens, depending on its exceptionality. The death of a child is definitely a 'fact', but in itself it does not become news. It becomes news when consensus is achieved over its significance.

In the journalist's professional examination, the information is certified both as to the indubitability of the 'fact' and its relevance and degree of public interest. Instead, communication is seen as the self-interested promotion by any social actor of a 'fact', often linked to the desire to persuade.

These prerogatives are lost if journalism's productive monopoly and, as a consequence, in part also the function of defining what is newsworthy are lost. Facts are assembled, distributed and used in a spatially and temporally defined set—the 6 o'clock news or the morning newspaper—less and less. They become a constant flow brought to our attention—especially through the Internet—by individuals linked together in ways that are no longer defined—or at least no longer solely defined—by journalistic mediation (Deuze and Witschge 2018).

As a consequence, it becomes more difficult to stabilize their meaning because, by involving a greater number of actors, the shape and substance of the process to negotiate these meanings changes.

The fragmentation of the distribution processes and the segmentation of the publics negatively affect the consensus, making more unstable and exposed to continuous upsets (Marini 2017) the role attributed by McLuhan to the media, namely, being at the origin of common sense (Gili and Maddalena 2017), meant as a dynamic process aimed at underpinning the social requirement to make comprehensible and familiar that which is not (Santambrogio 2006). Those information bubbles (Pariser 2011) in which we all enclose ourselves expand, those bubbles where we obtain reassurances on reality through narrations that are culturally closest to us (Sunstein 2001) get bigger.

What McNair (2006) defines as cultural chaos consists precisely of the messages' lacking a single direction, thrown into an overcrowded field of communication where lower entry barriers multiply the newsworthiness of events, subjects and topics, promoting a richer but also more confused public discourse (Bennett and Pfetsch 2018). At the same time, the timing of the production, distribution and use of this news merge more and more into de-spatialized instantaneousness (Thompson 1995).

The term chaos appears particularly effective because it contrasts with the concept of control, which is central to the institutionalization of journalism as a place for identifying what is significant and to be shared. Indeed, the terms most used to define journalism and journalists—respectively 'fourth estate' and 'gatekeeper'—are based on selection and simplification processes precisely for the purpose of control. Whether acting in agreement with the political and economic powers-that-be, perhaps in a manner underlining its subordination, or contrasting with them, according to the logics of watchdog journalism, journalism nevertheless decides what to safeguard and what to let go. It is control work which substantiates that hierarchical tendency recalled by Carlson (2017); however, to again take up McNair, it is destructured from the current chaos, which potentially allows everyone to enter the communication arena, not only as a consumer but also as a producer.

The single news items are no longer inserted in their context of signification thanks to journalistic contextualization, but also through the reading and interpretation on the part of the sources and the public. The user is not faced with a defined text— what is written in the newspaper or seen on TV—requiring his or her interpretation, perhaps by discussing it with fellow users; instead the text in question is in continual evolution thanks to the immediacy of comments, specifications and contestations, produced both by the sources of the news and other users. The news becomes an open resource, as shown by the continual references in journalistic pieces to a whole variety of different texts.

All users enter the journey along which a news item develops. This participatory tendency is pliably highlighted by the characteristics of social network communication: in the same setting we receive news from traditional mainstream journalism, but we also produce information, mainly aimed at the private circuit of our 'friends'; nevertheless, it lets us more immediately perceive the dialogical nature of the information process.

Journalism has historically built its legitimation on the capacity to define and standardize the process to select, verify and rank the facts to transform into news. Nevertheless, if the hegemony of a paradigm based on the lasting supremacy of journalistic procedures is fading (Nerone 2013), it becomes necessary to identify other bases upon which to conceive of journalistic authority and legitimize its pre-eminence in recounting reality. These bases must be able to support the loss of the exclusivity on which journalistic work was founded and the hierarchical restrictions mentioned earlier, owing to which the news to be made known was transmitted in a top-down manner. Now the expansion of newsworthiness, which makes the reconstructive nature of journalistic mediation clearer, but also reveals its bias and deliberate selectiveness, the destructuring and fragmentation of a hybrid journalistic system, as well as the horizontal forms of digital distribution give a centrality to other values underlying the communication pact such as participation and transparency (Carlson 2017).

However, participation and transparency have not yet been consolidated as unan-imously defined and accepted procedures, also—and perhaps above all—because they need the users to be definitively and consciously included in the information process. Such a cultural evolution is difficult to accept both for the journalists, who primarily see this process as a progressive devaluation of their professionalism, and for the public of users, who are not always able to make the most of all the dialogical opportunities provided by the digital technologies, or are not always interested in doing so.

It is not a simple evolution because it requires the legitimation of authority not to be intrinsic *to* the role, but instead to be played out each time *in* the role. That concept of negotiated authority recalled by Giddens (1990), according to which legitimate authority cannot be recognized merely due to covering a particular social role, but instead must be conquered in the relationship each time, also extends to journalism.

Authority becomes a dialogical resource that is only defined and consolidated thanks to the ability to call up several perspectives on every issue, recognizing each one's equal dignity. Thus, this highlights the characteristic of journalism as a public discursive space where it calls up and negotiates with its set of interlocutors. Signifi-cance is attributed through more complex cognitive journeys, because it results from the wide variety of proposals, models and values which appear on the social scene.

As is clear for all to see, it is a true revolution that requires a great capacity for inclusion and openness. However, for time being, this is a long way from being realized because we are still blocked in a transit phase in which defensive propen-sities prevail. The journalists, for their part, see their professional identity made uncertain both owing to the ever greater distrust in their work, and the difficulties of the traditional business models and the consequent weakening of their consolidated operating procedures. A similarly cautious attitude can also be seen on part of the citizens, vaguely aware of their actorship in a journalistic field undergoing great dif-ferentiation, but still inclined to seek reassurances, mainly guaranteed by enclosing themselves in their own points of view, relying on definitions of situations that they most agree with and that mitigate the strong sense of prevailing uncertainty. Forms of reconstructing reality in line with personal visions of the world, those spoken about by Cass Sunstein and Eli Pariser in their theories of echo chambers and information bubbles, continue to predominate.

References

Anderson CW, Bell E, Shirky C (2012) Post-industrial journalism: adapting to the present. Tow Center for Digital Journalism, New York

Bakardjieva KM (2010) The internet and subactivism: cultivating young citizenship in everyday life. In: Olsson T, Dahlgren P (eds) Young people, ICT's and democracy: theories, policies, identities and websites, Nordicom, Göteborg, pp 129–146

Bennett WL, Pfetsch B (2018) Rethinking political communication in a time of disrupted public spheres. J Commun 68(2):243–253

Benson R, Neveu E (2004) Bourdieu and the journalistic field. Polity Press, New York

Bolter J, Grusin R (1999) Remediation. Understanding new media. MIT Press, Cambridge, Massachusetts

Cappella J, Jamieson K (1997) Spiral of cynicism: the press and the public good. Oxford University Press, New York

Carey J (1989) Communication as culture. Routledge, London

Carlson M (2017) Journalistic authority legitimating news in the digital era. Columbia University Press, New York

Cartocci R (2007) Mappe del tesoro. Il Mulino, Bologna

Couldry N (2010) Why voice matters. Sage, London

Dahlgren P (2009) Media and political engagement. Citizens, communication and democracy. Cambridge University Press, New York

Deuze M (2012) Media life. Polity Press, Malden

Deuze M (2015) What is journalism? Professional identity and ideology of journalists reconsidere. Journalism 16(4):442–464

Deuze M, Bruns A, Neuberger C (2007) Preparing for an age of participatory new. Journal Pract 3(1):322–338

Deuze M, Witschge T (2018) Beyond journalism: theorizing the transformation of journalism. Journalism 19(2):165–181

Giddens A (1990) The consequences of modernity. Polity Press, Cambridge

Gili G, Maddalena M (2017) Chi ha paura della post-verità. Marietti, Genova

Hall S et al (1978) Policing the crisis: mugging, the state and law and order. Macmillan, London

Hanitzsch T, Vos TP (2018) Journalism beyond democracy: a new look into journalistic roles in political and everyday life. Journalism 19(2):146–164

Hjarvard S (2013) The mediatization of culture and society. Routledge, New York

Heinrich A (2011) Network journalism: journalistic practice in interactive spheres. Routledge, London

Kovach B, Rosenstiel T (2001) The elements of journalism: what news people should know and the public should expect. Three Rivers Press, New York

Knorr Cetina K (1981) The manufacture of knowledge. An essay on the constructivist and contextual nature of science. Pergamon, New York

Lorusso AM (2018) Postverità. Laterza, Bari-Roma

Mancini P (2013) Media fragmentation, party system and democracy. Int J Press/Polit 18(1):43–60

Marini R (2017) Potere dei media, interdipendenza tra poteri e pluralismo dell'informazione. Problemi dell'informazione 42(1):3–30

Marshall TH (1950) Citizenship and social class: and other essays. Cambridge University Press, Cambridge

McQuail D (2013) Journalism and society. Sage, London

McNair B (2006) Cultural chaos, journalism, news and power in a globalized world. Routledge, London

McNair B (2018) From control to chaos, and back again. Journal Stud 19(1):499–511

Nerone J (2013) The historical roots of the normative model of journalism. Journalism 14(4):446–458

Norris P (1999) Critical citizens: global support for democratic government. Oxford University Press, Oxford

Pariser E (2011) The filter bubble: what the internet is hiding from you. Penguin Press, New York

Peters C, Broesma M (2013) Rethinking journalism. Trust and participation in a transformed news landscape. Routledge, London

Pizzorno A (2007) Il velo della diversità. Feltrinelli, Milano

Putnam RD (2000) Bowling alone: the collapse and revival of American community. Simon and Schuster, New York

Riegert K (2007) Politicotainment: television's take on the real. Peter Lang Publishers, New York

Robinson S (2009) Journalism as process: the organizational implications of participatory online news. Journalism and Communication Monographs 13(2):137–210

Rodotà S (2013) Iperdemocrazia. Come cambia la sovranità democratica con il web, Laterza, Bari-Roma

Rosanvallon P (2006) La contro-democratie à l'age de la defiance. Seuil, Paris

Santambrogio A (2006) Il senso comune. Laterza, Bari-Roma

Schudson M (1978) Discovering the news. A social history of American newspaper. Basic Books, New York

Schudson M (1996) The good citizen. A history of American civic life. The Free Press, New York

Singer JB (2018) Transmission creep. Media effects theories and journalism studies in a digital era. Journal Stud 19(2):209–226

Sorrentino C (2006) Il campo giornalistico. Carocci, Roma

Sunstein C (2001) Echo chambers: Bush V. Gore, impeachment, and beyond. Princeton University Press, Princeton

Thompson M (1995) The media and modernity. a social theory of the media. Stanford University Press, Stanford

Tong J (2018) Journalistic legitimacy revisited digital. Journalism 6(2):256–273

Tuchman G (1973) Making news by doing work: routinizing the unexpected. Am J Sociol 79(1):110–131

Laura Solito Ph.D in Sociology at University of Pisa, she is Associate Professor in Sociologia della Comunicazione since 2007 at the University of Florence, Department of Social and Political Sciences. Since 2015 she has been Vice-President of Communication and Public Engagement in the University of Florence. Her main fields of research are on the role of communication in the public administration and especially in the production and management of services and on new forms of citizenship favored by digital communication.

Carlo Sorrentino Ph.D in Sociology at University of Florence, he is Full Professor in Sociologia dei processi culturali since 2011 at the University of Florence, Department of Social and Political Sciences. His main field of research is Journalism studies and in the last years he is investigating on the role of journalism in the trasformation of the public sphere. He is editor of the journal *Problemi dell'informazione* edited by Il Mulino.

From Meta-Journalism
and Post-Journalism to Total Journalism

Xosé López-García, Alba Silva-Rodríguez, Sabela Direito-Rebollal
and Jorge Vázquez-Herrero

Some evidence provided by the transition to the network society and the arrival of the disruption announced by the Internet of Things and the smart automation.

Abstract Journalism is facing challenges introduced by the network society with strengths and weaknesses. On the one hand, it is the legacy of more than a century and a half that offers multiple evidence of its social utility for receiving trustworthy news, generating knowledge, and achieving a better operation of plural and democratic societies. On the other hand, a loss of credibility, difficulties to show their contribution within a context dominated by confusion and unfounded rumours, and the multiplication of problems to guarantee sustainability of media investing in quality journalistic projects. But journalism is still alive and it remains at the heart of social debates within the present turbulent context. In order to guarantee its continuity in the digital era, characterized by disruptive changes and a communication metamorphosis in various stages, journalism is looking for its reinvention without sacrificing the essentials: telling, explaining and interpreting what is happening in society, using techniques and tools to guarantee veracity.

Keywords Total journalism · Digital media · Technology · Journalism

X. López-García · A. Silva-Rodríguez (✉) · S. Direito-Rebollal · J. Vázquez-Herrero
Universidade de Santiago de Compostela, Santiago de Compostela, Spain
e-mail: alba.silva@usc.es

X. López-García
e-mail: xose.lopez.garcia@usc.es

S. Direito-Rebollal
e-mail: sabela.direito@usc.es

J. Vázquez-Herrero
e-mail: jorge.vazquez@usc.es

© Springer Nature Switzerland AG 2020   199
J. Vázquez-Herrero et al. (eds.), *Journalistic Metamorphosis*,
Studies in Big Data 70, https://doi.org/10.1007/978-3-030-36315-4_15

1 Introduction

The history of journalism is full of lights. The Watergate was one of the most shocking investigations due to the relevance of the case and the consequences—the resignation of the president of the main world power—. Journalism has also its shadows, such as the fake news disseminated in the last decades of the twentieth century and the beginning of the twenty-first century by two reference media. *The Washington Post* published the report *Jimmy's World* by Janet Cooke, that won a Pulitzer and was withdrawn because the kid did not exist, and *The New York Times* published 36 fake articles written by one of its young figures, Jayson Blair. These were not the only cases: a photographer from *Los Angeles Times* retouched one of his photos from the Iraq War to add dramatism to the scene; the columnist Foster Williams, from *The Wall Street Journal*, was convicted for having sold privileged information. But maybe those were the most striking of journalism between the centuries, one of its golden times, which served to feed a permanent debate on the verification of news and to reflect on the limits between reality and fiction when telling stories and the compliance with the codes of conduct of journalism.

Long before, the Muckrakers—waste trackers—, called contemptuously in 1906 by the North-American president Theodore Roosevelt, wrote glorious pages of investigative journalism. At the end of the nineteenth century and the beginning of the twentieth century, a group of professionals prepared to tell stories about injustice and corruption started a needed path: the complaint (Campos González 2015). Their example was followed by other professionals that discovered and explained relevant facts, which were illustrated with the Watergate and the path opened by WikiLeaks, with documents' leaks containing reserved and relevant information for the public interest, exploited by journalists to explain transcendent stories.

The list of great investigative reporters has also proper names in Europe, among them Gunter Wallraff, 'the undesirable journalist', who lived one year as an immigrant in his own country, Germany, to write about abuses and the racist treat with Turkish. Also, the French Edwy Plenel produced research reports during the presidency of François Miterrand, talking about the role of the president and French secret services in the sinking of the Rainbow Warrior, the flagship of the ecologist organization Greenpeace, at the port of Auckland in New Zealand. Also, he discovered the hidden details of the bonds between Gadafi and Sarkozy during the presidency of the latter and also during the term of François Hollande. He discovered foreign undeclared accounts that resulted in the resignation of the minister Jerôme Cahyzac. Prominent positions are also occupied by the Italian Fabrizio Gatti—specialized in immigration topics—, the Spanish Xavier Vinader, Manuel Cerdán and Mar Cabra. In the last few years, networks of investigative journalists have emerged, such as the International Consortium of Investigative Journalists (ICIJ), a US-based non-profit global network which has the purpose to provide investigative journalists around the world with resources. A group of European journalists belong to the network.

Journalism has shown, throughout its history, its relevant role to feed plural and informed societies. It continues to be the most established and widespread way of society to generate and combine knowledge in all areas of life (Godler et al. 2018). However, its contributions do not prevent the multiplication of criticism due to its weaknesses and mistakes, and its presence in debated on stable and sustainable models of democratic and plural societies beyond the known standards in countries that have reached a greater development and a better rating of its own citizens and foreigners.

2   The First Great Turnover

Since the middle of the twentieth century, on the basis of experiences and within changing social contexts, journalism started a transformation process with different movements that introduced new narrative ways and renewed techniques and dimensions to produce information. The stronger movement started in the North-American journalism, which had positioned itself as dominant in many countries. But there were signs of change in the same direction from many countries—even before the emergence of a new journalism with the brand USA—, including the Argentinian Rodolfo Walsh with its *Operación masacre*, and the Spanish Manuel Chaves Nogales with *Juan Belmonte: matador de toros* and *A sangre y fuego*. The common thread was the application of resources from literature to the non-fiction story, with renovated ways within the old coexistence between journalism and literature throughout history (Cuartero Naranjo 2017).

In the second half of the twentieth century, it was generated a social, political and economic context which boosted the explosion of journalism. There was an important development of mass media in the majority of countries around us, and technologically-mediated communication acquired a main role in the functioning of societies (Mompart et al. 1999). This scenario and the situation in the USA, characterized by great cultural and social changes, led to the birth of the so-called 'new journalism', of which Tom Wolfe made the first anthology (Wolfe and Johnson 1973). The relation between journalism and literature was exploited by a group of North-American professionals—Norman Mailer, Truman Capote, Tom Wolfe, Gay Talese, Hunter S. Thompson, Joan Didion, among others—to tell present facts using literary resources, that is, in the domain of the factual word (Chillón 2014).

In the second half of the second millennium, another journalistic trend emerged in the United States: precision journalism. Due to the contributions from computing and databases, together with the methodological techniques provided by science, it advocated a new method for producing pieces of news. The search for techniques that allow journalists to offer stories containing more evidence and guaranteeing trustworthiness was a challenge for Professor Philip Meyer, who wrote the first essay on the new journalistic trend that created and nourished with his proposals (Meyer 1973). Precision journalism aims at producing pieces of news based on empirical data collected and verified using scientific methods of socio-statistics and computing research (Dader 1995). Computing entered in journalistic practices to support processes of narrative renewal and to provide more scientific soundness in the construction of stories than what happens in society.

Further, dimensions were also added, introducing the service journalism, civic journalism and solution journalism, among other trends. The result in the twentieth century was the inclusion of tools and the renovation of techniques that have oxygenated the professional practice and have provided vitamins to face challenges of market competence among the main media (press, radio and television), while responding to the demands of sectors of the citizenry.

3   The Digital Revolution

The turn of the century between the second and the third millennium has been marked by relevant debates in the profession, while the Internet advanced in its development with the Web and the digital transition of the media was entering the network society. We refer to a relatively quickly process. In March 2014, 25 years after Tim Berners-Lee drafted the document that defined the World Wide Web and the hypertext protocol (HTTP), most of printed media were already online, transferring contents and exploring possibilities without having clear goals. Thanks to the Information and Communication Technologies (ICT) and innovation programs, both the media and the news uses and consumption initiated a quick and permanent reconfiguration process that radically changed the ecosystem in communication.

Legacy and digital native media opened up an own path where hypertextuality, multimediality and interactivity emerged as distinctive and characteristic features of the new online media or cybermedia. This created the conditions for the emergence of cyber journalism as an area of specialization in journalism that uses cyberspace to investigate, produce and, above all, spread news contents (Salaverría 2005). Despite the so-called crisis of the 'dotcom' in the beginning of the third millennium, in the second decade of the twenty first century, online media redefined their strategies while the Web 2.0 was emerging. It refers to the second generation of the Internet, with the social Web and an extensive list of resources and practices to help users socialize contents and explore options for active audiences.

The adaption and transformation process from traditional media to the new digital formats and interfaces, "far from constructing a simple technical adjustment, has contributed to a gradual transformation of the media and their audiences" (Peña-Fernández et al. 2016). Within that complex and diverse transformation process, changes and renovations of narrative techniques occurred, using old and new tools. The result has been, despite the economic crisis in 2008 and the adaption problems and transformations of the media, a breeding ground for innovation in journalism, in the search for new ways to tell stories, to build user confidence in the news and eventually to define an own space within a context full of noise and fake news.

4 The Second Great Turnover

The digital impact on journalism encouraged answers to challenges brought by current technologies and their social appropriation by citizens. The changing scenario, both in society and in the communication arena, motivated the emergence of a second great renovation. It was digital and it was fed by dimensions provided by daily creativity and the models sponsored by large technology companies and device manufacturers. In the midst of dramatic changes, and without much time to reflect, new formats and expressive modalities to tell non-fiction stories arose.

Data-based techniques, virtual reality and expanded stories using different means and channels entered into force and initiated paths of model exploration and consolidation. Data journalism is being implemented in the media with own space (Ferreras Rodríguez 2014). This trend is complemented with news pieces made by specific teams created to encourage this professional practice. The growing relevance of Big Data in the network society has opened up a new path that journalism professionals, with interdisciplinary teams, face with good results in the media all over the world. Although the main experiences are in North-American media, today it is a consolidated trend among the world top media.

The boom of the virtual reality, with relevant campaigns for the introduction on the market of devices at affordable prices, has also reached journalism, where immersive experiences are currently a trend in the main media outlets. Immersive journalism, whose narratives are based on 360 recording techniques, virtual reality and 3D enable the receiver to feel closer to the news scenario, as he experiences a spatial and sensorial immersion that allows him to feel 'being there'—also known as presence—(Pérez Seijo and López-García 2017). Although these techniques pose ethical challenges to professionals, their contributions have found a place in some of the major media outlets and there are relevant labs that encouraged them, both in media and technological companies.

Under the wing of convergence and the evolution of the network society, the transmedia model, with its techniques and expanded narratives, encouraged a new dimension for journalism. The production of fractionated news pieces for their dissemination through multiple platforms using extended stories has become a widespread method to face challenges in the media ecosystem of the third millennium. These narratives have proven themselves to be a successful formula to bring the creative effort of different contents together, to generate involvement of the fandom and to increase channels to make the project profitable (Costa 2013). While its main success lies in fiction narratives, transmedia storytelling has initiated a journey in non-fiction stories and the media are progressively testing experiences.

Online media find in interactive non-fiction an appropriate place for the development of stories and, little by little, adopt new interactive and hybrid ways to face the challenge of the Internet and to meet the needs of more active and permanently connected audiences (Vázquez-Herrero et al. 2017). All the available data show this is a proper path to follow, although there are a lot of dimensions to explore and there is a need of renewed formulas to improve efficiency in communication and connection with users.

Legacy from the past and new experiences advised the creation of labs to experiment models and formats in order to provide more added value to users and to successfully compete in the network society, where spaces are highly contested. The labs were consolidated as a way for innovation in the media, with different models to face an adverse scenario (Salaverría 2015). This challenge has created an excellent breeding ground to convert labs into an emerging trend for cybermedia and the area of communication.

These labs, most of them established as of 2010, work in innovation. Meanwhile, two projects under construction in the scenario of the network society encourage new dimensions for communication and journalism: the Internet of Things and the smart automation. When the 5G mobile technology knocks at the door, transformations in the communication ecosystem will progress with greater strength and offering symptoms of the conquest of unexplored territories.

5 Total Journalism

Renovations experienced by journalism in the second and the third millennium have brought assets for providing more added value and more benefits for citizens. In order to take advantage of all those contributions and those to come from labs created by many media to encourage innovation, the Internet of Things and the smart automation, it is necessary to design *total journalism* projects, a name chosen to put the spotlight on the reinvention of the journalism and the result of transformation processes in the years since the early 2000s. In other words, we should understand total journalism as that renewed practice which expresses using multimediality, hypertextuality and interactivity, traditional journalism—raw—adapted to the present society, which uses data, immersive and transmedia techniques, among others, and which guarantees the truthfulness of information and the performance on differential values for users.

From this conceptual approach, when analysing the future of journalism and the future journalism, we need to answer basic questions. Firstly, what are the elements that remain from traditional journalism, modern journalism—cultivated for more than a century and a half—and inherited journalism? Secondly, what has changed in journalistic precepts, in processes of social institutionalization, in citizens' participation—in processes of news-contents co-elaboration—, in social precepts and in the own conception of journalists?

As a starting point and in schematic form, with the data available from the analysis of current journalistic trends and from the result of the two great and recent renovations, referred in the first part of this chapter, we can affirm that the basic elements of journalism have not changed. Contemporary journalism is a social communication technique that bases its activity in the production of trustworthy and useful information for society, using different techniques and platforms. Its main goal is to provide reliable information—so verification is one of the essential points (Kovach and Rosenstiel 2007)—which citizens need to intervene in society, in social, political and economic processes.

It belongs to society and it is within it; thus, it changes with society too. And, in the new media environment, where most of social players and a lot of citizens produce contents, it has to offer pieces with added value.

However, there are also moves. What has changed? There are multiple evidence of the metamorphosis of journalism in the last decades. Journalism has experienced constant transformations throughout the history and these have been accelerated since the popularization of the Internet, the Web, the social Web and the development of current technologies. Tools have changed, but also the way we use and consume news within a context of overabundance of information. Connectivity and mobility have created a new scenario, with a renewed communicative ecosystem, updated techniques and many challenges. The transition of traditional media has not ended, and digital native media have tried to break rules to get a foothold in the contemporary ecosystem.

Changes in the network society led to changes in the social perception on the role of journalists and the media, after successive processes of loss of credibility. Unfounded rumours and all we group around the concept of post-truth, are the finishing touch to this discredit, which is also an opportunity. Hence, tools, techniques, social perception and the own vision of journalists on their role have changed.

6   In Conclusion

Specialization has provided a lot of answers to challenges in last years, both as regards topics, technology and narrative. For this reason, as a result of advances and evidence provided by journalism's innovation labs—technological, communication companies and so on—, the establishment of integrated packages for the journalism practice is imposed.

In order to apply integrated packages in the training of journalists, in works on media literacy—educommunication—and in research itself, they should contain attention to changes in mobile and connected society, attention to new processes of uses and consumptions—channels and themes—, attention to the renovation of techniques, etc.

However, which practices should be prioritized nowadays? We should mention: data journalism, immersive journalism, multimedia and multi-format narratives, transmedia narratives, verification techniques (fact-checking) and semi-automated systems. As well as we should include: resources to respond to new challenges through the creation of international research networks, the creation of specialized networks and collaboration mechanisms, encouragement of journalism's innovation labs and educommunication programs.

Within this context, in the 2020s, we should go beyond occasional names to analyze changes, such as meta-journalism—writing about what we do beyond/after journalism—and post-journalism—which assumed the role of journalism from the twentieth century, after journalism from the industrial era—. The conceptual turn goes towards total journalism as a way to bring together the inherited—perceptive, reflections and research on that perceptive and professional practice—, things that emerged in that first transition stage to the digital scenario, after that industrial journalism, and the knowledge provided by the scientific community during decades. Thus, journalism could be able to offer a set of techniques, within the frame of various trends, movements and ways to understand journalism, more solid and efficient to produce trustworthy pieces of news.

The proper use of accumulated scientific knowledge in the field and the application of renewed techniques, in the successive historical stages and specially in the last stage, are an excellent antidote for journalism to offer quality information and to contribute to fight for misinformation that is taking possession of renewed spaces of the communicative ubiquity in the network society. That is, we should apply total journalism within a sustainable context, and then we will contribute to a better-informed society.

Acknowledgements This article has been developed within the research project *Digital native media in Spain: storytelling formats and mobile strategy* (RTI2018-093346-B-C33) funded by Ministry of Economy and Competitiveness (Government of Spain) and co-funded by the ERDF structural fund, as well as it is part of the activities promoted by Novos Medios research group (ED431B 2017/48), supported by Xunta de Galicia. The author Sabela Direito-Rebollal is a beneficiary of the Faculty Training Program funded by the Ministry of Science, Universities and Innovation (FPU15/02557), as well as Jorge Vázquez-Herrero (FPU15/00334).

References

Campos González V (2015) ¡Extra, extra!: Muckrakers, orígenes del periodismo de denuncia. Ariel, Barcelona

Costa C (2013) Narrativas Transmedia Nativas: Ventajas, elementos de la planificación de un proyecto audiovisual transmedia y estudio de caso. Historia y Comunicación Social 18:561–574

Cuartero Naranjo A (2017) El concepto de Nuevo Periodismo y su encaje en las prácticas periodísticas narrativas en España. Doxa Comunicación 25:43–62

Chillón A (2014) La palabra fáctica. Literatura, periodismo y comunicación. Colección Aldea Global, Universidad Autónoma de Barcelona, Universitat Jaume I, Universitat Pompeu Fabra, Universitat de València, Barcelona

Dader JL (1995) Periodismo de precisión: la observación matemática de la actualidad. In: El-Mir AJ, Balbuena F (eds) Manual de periodismo. Universidad de Las Palmas, Prensa Ibérica, Las Palmas

Ferreras Rodríguez EM (2014) El periodismo de datos en España. Estudios sobre el Mensaje Periodístico 22(1):255–272

Godler Y, Reich Z, Miller B (2018) Social epistemology as a new paradigm for journalism and media studies. New Media and Society [Online first, 20 December]

Mompart Gómez, Marín JLl, Otto E (1999) La irrupción de la información televisiva y la influencia del periodismo singular. In: Mompart Gómez, Marín Otto E JLl (eds) Historia del periodismo universal. Editorial Síntesis, Madrid

Kovach B, Rosenstiel T (2007) The elements of journalism. Three Rivers Press, New York

Meyer P (1973) The precisión journalism. Indiana University Press, Bloomington

Peña-Fernández S, Lazkano-Arrillaga I, García-González D (2016) La transición digital de los diarios europeos: nuevos productos y nuevas audiencias. Comunicar 46(24):27–36

Pérez Seijo S, López-García X (2017) Periodismo inmersivo y radiotelevisiones públicas europeas: comparativa de las aplicaciones móviles de consumo inmersivo de no ficción. In: Herrero J, Trenta M (eds) El fin de un modelo de política. Cuadernos Artesanos de Comunicación 129. Sociedad Latina de Comunicación Social, Tenerife

Salaverría R (2005) Redacción periodística en Internet. EUNSA, Pamplona

Salaverría R (2015) Los labs como fórmula de innovación en los medios. El Profesional de la Información 24(4):397–404

Vázquez-Herrero J, Negreira-Rey M-C, Pereira-Fariña X (2017) Contribuciones del documental interactivo a la renovación de las narrativas periodísticas: realidades y desafíos. Revista Latina de Comunicación Social 72:397–414. https://doi.org/10.4185/RLCS-2017-1171

Wolfe T, Johnson EW (1973) The new journalism. Harper and Row, New York

Xosé López-García Professor of Journalism at Universidade de Santiago de Compostela (USC), Ph.D in History and Journalism (USC). He coordinates the Novos Medios research group. Among his research lines there is the study of digital and printed media, analysis of the impact of technology in mediated communication, analysis of the performance of cultural industries, and the combined strategy of printed and online products in the society of knowledge.

Alba Silva-Rodríguez Associate Professor of Journalism at the Department of Communication Sciences at Universidade de Santiago de Compostela. She is Ph.D. in Journalism and Member of Novos Medios research group. She is secretary of the RAEIC journal. As a researcher she focuses on the assessment of digital communication, especially the study of mediated conversation in social media and the evolution of media contents in mobiles devices.

Sabela Direito-Rebollal Ph.D. Candidate at Universidade de Santiago de Compostela (USC). She holds a Degree in Audiovisual Communication (USC), a Master's Degree in Communication and Creative Industries (USC) and a Degree in Movie and TV Script from the Madrid Film Institute. She was visiting scholar at the University of Hull (United Kingdom) and the Vrije Universiteit Brussel (Belgium). Her research focuses on innovation, audience trends and programming strategies of European public service media.

Jorge Vázquez-Herrero Ph.D. in Communication, Universidade de Santiago de Compostela (USC). He is a Member of Novos Medios research group (USC) and the Latin-American Chair of Transmedia Narratives (ICLA–UNR, Argentina). He was visiting scholar at Universidad Nacional de Rosario, Universidade do Minho, University of Leeds and Tampere University. His research focuses on digital interactive non-fiction digital storytelling—mainly interactive documentary—, micro-formats and transmedia, immersive and interactive narratives in online media.

Printed in the United States
By Bookmasters